INVESTIGATIONS
for GCSE Biology

P W Freeland

BSc MPhil DipEd FIBiol
Head of Science, Worth School, Sussex

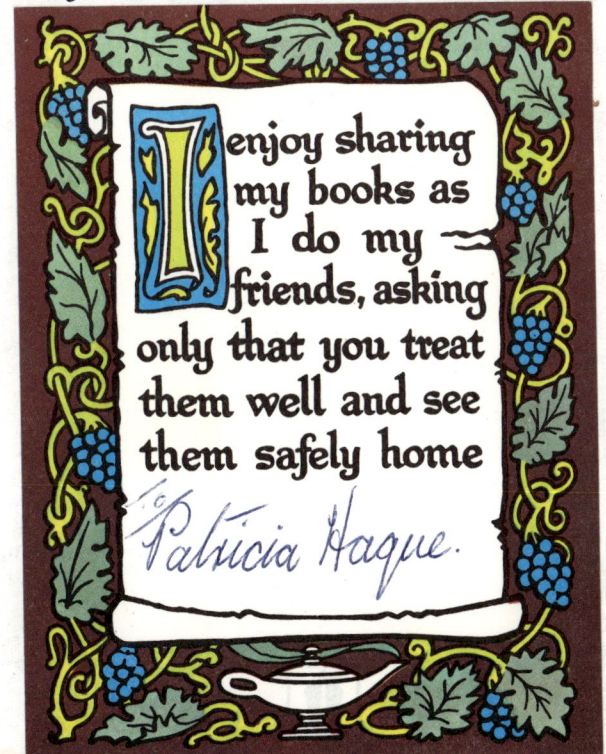

HODDER AND STOUGHTON
LONDON SYDNEY AUCKLAND TORONTO

Preface

This book is written for pupils in the 14–16 age range, particularly those working towards GCSE examinations in biology. It contains a large selection of practical investigations, designed to encourage an experimental approach to the subject. Although some investigations may be familiar, others involve new activities, both in the laboratory and in natural habitats. Each investigation involves carrying out a number of tasks, some simple and straightforward, others more difficult. In addition to encouraging observation and description of materials, the investigations provide opportunities for putting a wide range of skills into practice. You are asked not only to carry out set experiments, but to design your own methods, name your apparatus, select suitable controls, make calculations, draw conclusions and formulate hypotheses. Instructions are given on how to improve your drawings, and construct an artificial key for the identification of organisms.

Each of the investigations begins with background reading and a list of materials. This may be followed by a preparation, in which you set up apparatus, and one of observation, for taking measurements and recording data. **Bold type** is used in those instructions for which marks are awarded. When you have made your investigation, the skills you have used are assessed. The assessment tests your understanding of the topic, and ability to carry out an experiment.

Subjects for the investigations have been chosen to show that biology is a useful science. Many organisms produce useful products. Yeast, for example, may be used to produce bread and wine, while earthworms can be used to make compost. Making new discoveries and finding a use for them isn't difficult as long as you know how to go about it. Working through this book can help, as it suggests experiments for you to attempt.

Acknowledgments

Investigation 40 is based, with permission, on an investigation devised by D. G. Mackean. Figure 59 is based on a figure in *Plants and Mineral Salts* by J. F. Sutcliffe and D. A. Baker in the Studies in Biology Series published by Edward Arnold. The photographs reproduced in this book were supplied by Biophoto Associates (page 69), Colorsport (page 63), Copella Fruit Juices Ltd (page 26), Geoscience Features (pages 18, 43, 49, 51, and 61) and Science Photo Library (page 29).

Freeland, P.W.
   Investigations for GCSE biology pupils.
   1. Biology—Experiments
   I. Title
   574′.0724   QH316.5

ISBN 0 340 40527 9

First published 1987

Copyright © 1987 P W Freeland

All rights reserved. No part of this publication may be reproduced or transmitted in any form or by any means, electronic or mechanical, including photocopy, recording, or any information storage and retrieval system, without permission in writing from the publisher or under licence from the Copyright Licensing Agency Limited. Further details of such licences (for reprographic reproduction) may be obtained from the Copyright Licensing Agency Limited, of Ridgemount St, London WC1E 7AE.

Typeset by Macmillan (India) Ltd, Bangalore 25.
Printed in Great Britain for
Hodder and Stoughton Educational
a division of Hodder and Stoughton Ltd, Mill Road
Dunton Green, Sevenoaks, Kent by Page Bros (Norwich) Ltd.

Contents

1. Drawing specimens — 4
2. Making and using keys — 6
3. Testing for reducing sugars — 8
4. Reducing sugars, starch and proteins in foods — 10
5. Estimation of vitamin C — 11
6. Sugar and salt in food — 14
7. Spoilage of food by micro-organisms — 16
8. The souring of milk — 18
9. Environmental conditions affecting the digestion of starch — 20
10. The effect of different temperatures on trypsin activity — 22
11. Measuring enzyme activity by gas production — 24
12. Using enzymes to extract fruit juice: an introduction to biotechnology — 25
13. Enzyme inhibition by drugs — 27
14. Sugar, bacteria and tooth decay — 28
15. Demonstrating and making use of diffusion — 30
16. Diffusion of water through natural and artificial membranes — 32
17. Water loss from leaves — 34
18. Osmosis in living tissue — 36
19. Using a simple potometer — 39
20. Factors affecting the rate of photosynthesis — 40
21. Gas production by Canadian pondweed during photosynthesis — 42
22. Visits by bees to flowers — 44
23. Floral structure of tulip and polyanthus — 46
24. Wind as an agent of pollination and seed dispersal — 47
25. Distribution of seeds around a tree — 48
26. Depth of sowing and its effects on germination — 50
27. Producing new plants from stem cuttings — 52
28. Comparing rates of growth in parts of a seedling — 54
29. Tropic responses and auxin production — 56
30. Differences between seeds and seedlings — 58
31. Adaptation: form and function in plant stems — 59
32. Behaviour of the woodlouse — 60
33. Body size in mammals — 62
34. Exercise: measuring performance — 64
35. Carbon dioxide in inspired and expired air — 66
36. Fermentation of glucose by yeast — 68
37. Making use of yeast — 70
38. The effects of a selective weedkiller on a lawn — 72
39. The effects of artificial fertiliser on a population of water plants — 74
40. Some characteristics of a population of trees — 76
41. Finding the size of an animal population — 78
42. Pollution of water and air — 80
43. Organic pollution of water samples — 82
44. Water retention by sand, loam and peat soils — 84
45. Soil pH — 86
46. Organic matter in soil — 88
47. Decomposers in the soil — 90
48. Gene recombination — 92
49. The sense of taste — 94
50. The skin as a sense organ — 95

1 Drawing specimens

TIME
Investigation: 30–40 minutes
April–October

Drawing a biological specimen is one way of describing its structure. Some people find this quite difficult, as it usually involves reducing a three-dimensional object to a two-dimensional drawing. Always look very carefully at the outline of the object. Next, take a fairly hard pencil, such as a 2H, and sketch a rough outline on your paper, concentrating on shape and proportion. When you are satisfied that the outline is a fairly good likeness, make a firm outline, without any rough or blurred edges. Begin to add fine details to the drawing, but don't shade. It is better to make use of dots or short pencil strokes to show the effect of light, or the differences between rough and smooth surfaces. Finally, label parts of the drawing. Use a ruler to draw lines from the labels to different parts of the object. Space the labels so that none of the lines cross. See Figure 1.

It often happens that drawings have to be made larger or smaller than the specimen. Therefore, indicating the scale of a drawing is important. Scale drawings always preserve the correct proportions of a specimen, as in Figure 2. Increasing the scale of a drawing by a factor of four means that it is four times as long and four times as wide as the object you are drawing.

In this investigation you observe and make scale drawings of three different specimens.

Fig. 1 *Stages in drawing an acorn*

1 Use a hard pencil to sketch a faint outline. Concentrate on shape, size and proportion.

2 Make a firm, smooth pencil line to define the outlines of the specimen.

3 Add detail. Use short pencil lines and dots to suggest differences in texture, lighting, etc.

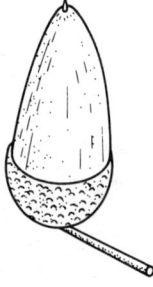

4 Add labels. Space the labels so that the lines do not cross.

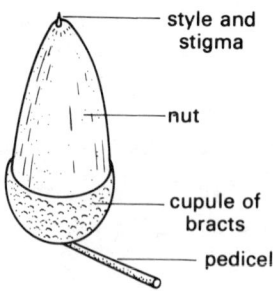

Fig. 2 *Scale drawings of the outline of a leaf*

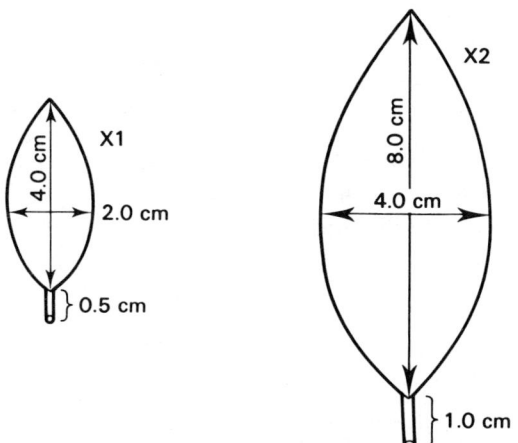
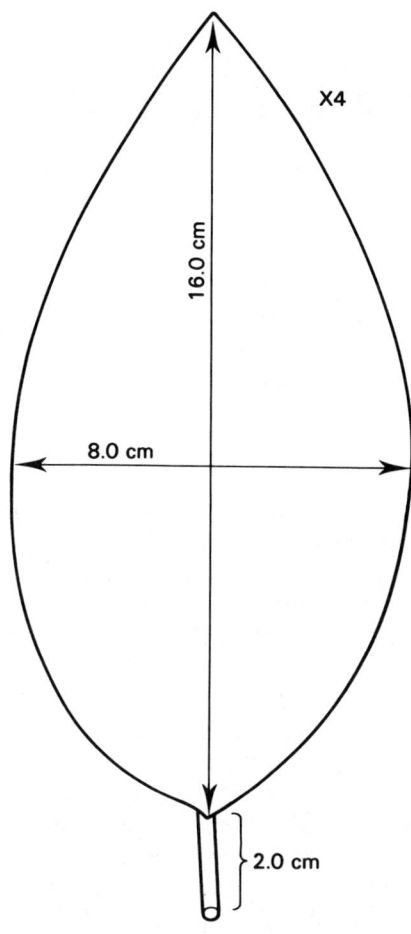

Investigation

Materials

- Pear
- Leaf or horse chestnut
- Preserved locust
- Scissors
- Forceps
- Hand lens
- Ruler, graduated in millimetres

Method

Part A

1 **Make a drawing, ×1, of the pear. Label the following regions on your drawing: pedicel (flower stalk); remains of flower.** (4)

Part B

2 **Draw the leaf of horse chestnut ×$\frac{1}{2}$, or ×$\frac{1}{4}$.** (6)
Show only the main veins in each leaflet. Label the following parts on your drawing: petiole (leaf stalk); leaflet, serrated margin (edge of leaflet). (6)

Part C

3 Using the forceps and scissors, remove the wings from the locust. **Make a large, labelled drawing, ×3, to show the ventral surface of the insect. Label any five of the following features on your drawing: head, thorax, abdomen, antenna, mandibles (jaws), femur, anus.** (10)

Total 20 marks

Taking it further

1 Collect a leaf of beech, privet, or daisy. Put it at the centre of a piece of graph paper and trace around its margin, to make a drawing of natural size. Find a method of making accurate scale drawings of the leaf, ×2 and ×4.

Use your method to draw enlarged outlines of the leaf (×2 and ×4) on the graph paper. In not more than five sentences describe the method you used.

2 Collect two large flowers from different garden plants. Make drawings to show the external appearance of each flower, then cut each flower in half and draw a vertical section.

3 Make drawings to show differences in structure between bones of a small mammal taken from (a) the fore-limbs and hind-limbs, and (b) different regions of the vertebral column.

2 Making and using keys

TIME
Investigation: 30–40 minutes

Biologists make keys so that people who are unfamiliar with plants and animals can identify them. When you make a key you need to know how each organism differs from others in a group. It is best to start with familiar non-living things, such as coins or stamps. Suppose you are asked to make a key for the identification of British coins and bank notes to the value of £10 (Figure 3(i)). You would start by collecting together all of the coins (1p, 2p, 5p, 10p, 20p, 50p, and £1) and bank notes (£5, £10) that were to feature in the key. Next, you would lay out the money, and take a close look at each coin and note (3(i)(a)). After that, you would try to sub-divide the coins and notes into smaller groups, each with one or more similar features (3(i)(b)). This process would continue until you were left with only one or two coins or notes in each group (3(i)(c) and (d)). When the grouping is complete, you can begin to write the key. Each number on the left of the key offers a choice between two or more features. Make your choice. The number on the right of the key then leads to your next choice, and so on, until the specimen is identified.

The aim of this investigation is to make and use artificial keys. Once you understand how to group objects such as coins, you should find it easier to make keys for identifying plants and animals.

Investigation

Materials

- 2p, 5p, and 10p coins
- Preserved locust, labelled A
- Defrosted prawn, labelled B
- Leaf of horse chestnut, labelled C
- Leaf of hazel, labelled D
- Ruler, graduated in millimetres

Fig. 3 *Making a key for the identification of British coins and bank notes*
(i) Dividing the money into groups

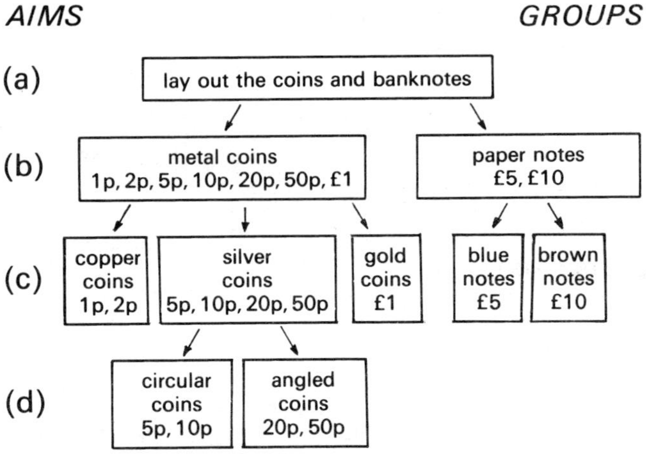

AIMS   GROUPS

(a) lay out the coins and banknotes

(b) metal coins 1p, 2p, 5p, 10p, 20p, 50p, £1 / paper notes £5, £10

(c) copper coins 1p, 2p / silver coins 5p, 10p, 20p, 50p / gold coins £1 / blue notes £5 / brown notes £10

(d) circular coins 5p, 10p / angled coins 20p, 50p

(ii) The completed key

(When making keys for the identification of plants and animals, features such as size and colour should be omitted, or used only when no other differences exist between organisms.)

1	Bank note	2
	Metal coin	3
2	Blue note, about 7.5 × 14.0 cm, with portrait of Duke of Wellington	£5
	Brown-orange note, about 8.5 × 15.0 cm, with portrait of F. Nightingale	£10
3	'Copper' coin	4
	'Silver' coin	5
	'Gold' coin, about 2 cm in diameter, with lion and unicorn, thistle, or leek design	£1
4	Coin about 2 cm in diameter, with a crown, chain and portcullis design	1p
	Coin about 3 cm in diameter, with a plumed feather design	2p
5	Coin with circular edge	6
	Coin with angled edge	7
6	Coin about 2 cm in diameter, with a crown and thistle, or crown and shield, design	5p
	Coin about 3 cm in diameter, with a rose and thistle design	10p
7	Coin about 2 cm in diameter, with a rose and crown design	20p
	Coin about 3 cm in diameter, with a Britannia design	50p

Method

Part A

1 Use the key in Figure 3 (ii) to identify the 2p, 5p, and 10p coins.

(a) **Write down, in the same order as they appear in the key, the features of the 2p coin by which it was identified.** (2)

(b) **Similarly, write down the features of the 5p coin by which it was identified.** (2)

(c) **By what features did you identify the 10p coin?** (2)

Part B

2 Examine the locust (A) and prawn (B). Use the key below to find the group to which the locust and prawn belong.

```
1  Bilaterally symmetrical              2
   Radially symmetrical          Coelenterates
2  Body segmented                        3
   Body unsegmented                  Molluscs
3  Body covered by an exoskeleton        4
   No exoskeleton                    Annelids
4  One pair of antennae               Insects
   Two pairs of antennae          Crustaceans
5  Three pairs of legs                Insects
   Four pairs of legs                Arachnids
   More than four pairs of legs    Myriapods
```

(a) **To which group in the key does the locust belong?** (1)

(b) **List in the same order as in the key, those features that led you to place the locust in this group.** (2)

(c) **To which group does the prawn belong?** (1)

(d) **List, in the same order, as in the key, those features that led you to place the prawn in this group.** (2)

Part C

3 You have a leaf of horse chestnut (C) and a leaf of hazel (D). The leaf of horse chestnut has five or more leaflets. It is a type of compound leaf known as palmate, because it looks like a hand. The leaf of hazel is simple and ovate.

Make a key to identify all of the leaves shown in Figure 4, and leaves labelled C and D. (8)

Total 20 marks

Fig. 4 *Five different leaves*

(i) Simple leaves – with an intact leaf blade

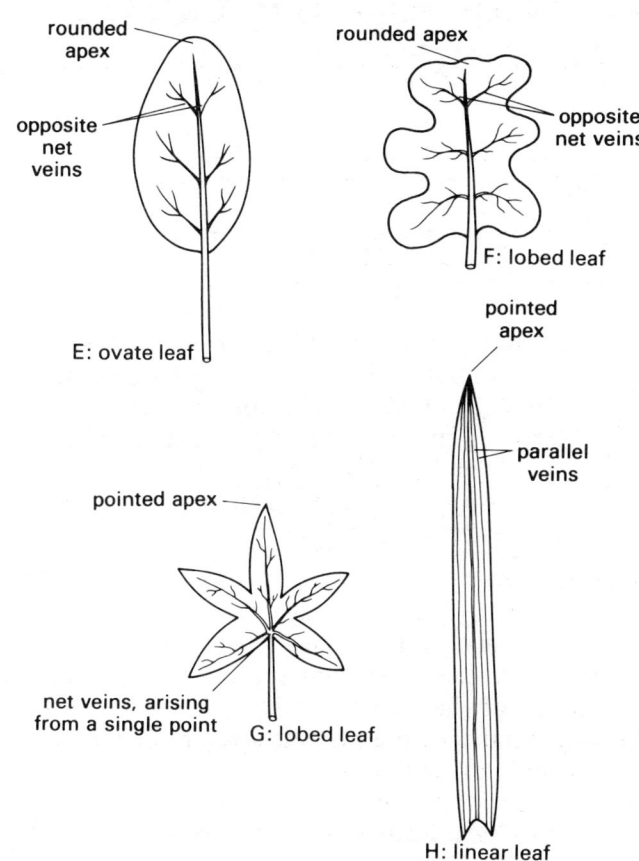

(ii) A compound leaf – with the leaf blade divided into separate leaflets

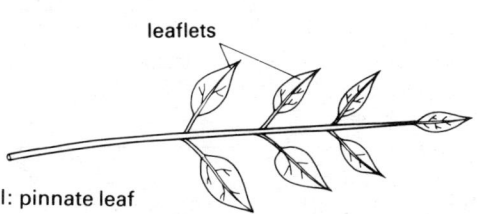

Taking it further

1 Make a collection of these items: pen, handkerchief, pencil, paper-tissue, watch, ruler. Divide the items into three pairs, each pair with one or more similar features. What are the similarities between items in each pair? What are the differences between items in each pair? Label the items from A–F and make a key to identify them.

2 Collect (a) flowers and (b) twigs from six different, known plants. Make a key that could be used by someone else to identify the plants.

3 Testing for reducing sugars

TIME

Investigation: 40–50 minutes

Three chemical reagents: Clinistix reagent strips, Benedict's solution and Clinitest tablets, can be used to test for sugars. Glucose, fructose and maltose are called reducing sugars, because they act as reducing agents in chemical reactions. When one of the reagents is reduced by a sugar, it changes colour. The tests are carried out as follows:

Clinistix test
Dip a reagent rtrip into a solution of the sugar, or crush plant-material against the pink square. If the square turns blue, a reducing sugar is present. Read the colour of the paper roughly ten seconds after dipping. Find the approximate amount of reducing sugar (glucose) from Table 1.

Table 1 *Concentrations of glucose shown by Clinistix reagent strips*

Colour of strip	Mass glucose (g) per 100 cm^3 solution
Negative (pink)	0
Light (red-purple)	0.25
Medium (purple)	0.5
Dark (blue-purple)	0.75

Benedict's test
Boil the reagent with a solution of the sugar, or a small piece of plant tissue. When the mixture boils it may become cloudy, changing colour from blue to green-brown. If the green-brown colour remains after cooling, a reducing sugar is present.

Clinitest
Put a tablet in a dry tube and add 0.7 cm^3 solution of sugar. After the mixture has cooled, read off approximate concentrations of reducing sugars from Table 2 or the manufacturers colour chart.

A fourth sugar, sucrose or cane sugar, is non-reducing. It does not produce a colour change with any of these reagents. To test for sucrose, the solution or tissue must first be boiled with an acid, or treated with an enzyme called invertase. This treatment hydrolyses sucrose into a mixture of glucose and fructose molecules. A positive test with any of the reagents, given only after heating with an acid, or adding invertase, shows that sucrose is present.

In this investigation you are going to test for reducing sugars and compare rates at which sucrose is converted into reducing sugars by (a) dilute hydrochloric acid and (b) invertase.

Table 2 *Concentrations of glucose shown by Clinitest tablets*

Colour of reagent	Mass glucose (g) per 100 cm^3 solution
Blue	0.0
Blue-green	0.25
Green	0.5
Green-brown	0.75
Brown	1.0
Yellow-brown	2.0 or more

Investigation

Materials
- Sugars A, B, C, and D
- Cubes of potato (X), onion (Y) and apple (Z)
- Four boiling tubes in a rack
- 20 cm^3 sucrose solution
- 1 cm^3 dilute hydrochloric acid
- 1 cm^3 invertase concentrate
- Benedict's solution
- Six Clinistix reagent strips
- Eight Clinitest tablets
- Clinitest colour chart
- Two 1 cm^3 plastic syringes
- 10 cm^3 plastic syringe
- Ten flat-bottomed tubes
- Test-tube holder
- Forceps
- Spatula
- Bunsen burner
- Glass-marking pen
- Safety spectacles
- Plastic gloves
- Graph paper

Method

SAFETY PRECAUTIONS

Put on safety spectacles and plastic gloves. Wash off any spillages with plenty of water.

Part A

1. You have four sugars labelled A, B, C, and D. Label the four boiling tubes A–D. Put one sugar into each of the tubes. Add 5 cm³ water to each tube, and shake gently until the sugars have dissolved. Add 2 cm³ Benedict's solution to each tube.

2. Light the bunsen burner. Take each tube in turn, hold it in the test-tube holder, and apply very gentle heat until the mixture boils. (*CAUTION*: avoid rapid heating, as this may cause 'bumping') After heating, return each boiling tube to the rack, and wait until it has cooled.

(a) **Tabulate your results.** (*4*)

(b) One of the sugars is sucrose. **Which one is it most likely to be?** (*1*)

(c) **Give a reason for your answer.** (*1*)

(d) **What further test is needed to confirm that this sugar is sucrose?** (*2*)

Part B

3. Use the Clinistix reagent strips to find out if reducing sugars are present in the potato (X), onion (Y) and apple (Z). Read the colour of the paper strip 10 seconds after crushing material against the pink square. Find the approximate amounts of reducing sugar in each material by referring to Table 1. **Tabulate your results.** (*4*)

Part C

4. Place eight flat-bottomed tubes on the bench, arranged in two rows of four. Use forceps to put a Clinitest tablet into each tube.

5. Label one of your remaining flat-bottomed tubes 'hydrochloric acid', and the other 'invertase'. Put the hydrochloric acid and invertase into the labelled tubes. Using a 10 cm³ syringe, add 10 cm³ sucrose solution to each tube. Place a 1 cm³ syringe into each tube, and use it to gently stir the mixtures.

Immediately, and at intervals of 5 minutes, over a period of 15 minutes, take 0.7 cm³ mixture from each tube and add it to a Clinitest tablet in one of the flat-bottomed tubes. See Figure 5. When the tablet has stopped fizzing, compare the colour produced with the colour chart. Record amounts of reducing sugar present.

(a) **Draw bar graphs of your results.** (*6*)

(b) **Which treatment has produced most reducing sugar?** (*2*)

Total 20 marks

Fig. 5 *Using Clinitest tablets to test for reducing sugars*

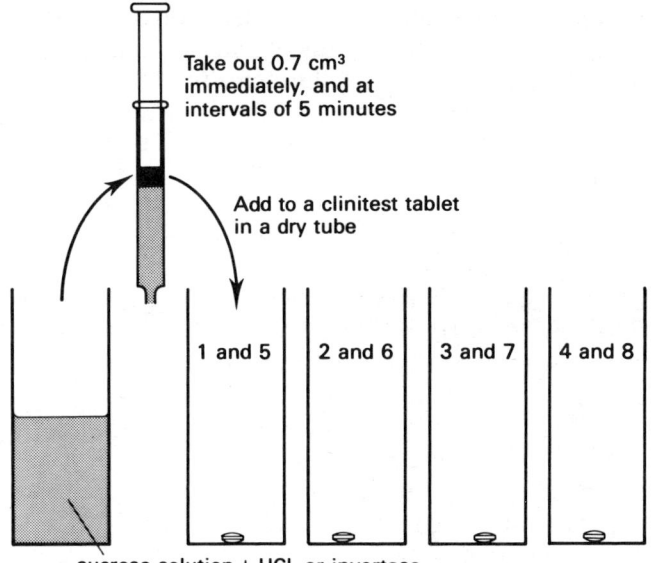

Taking it further

1. Cane sugar (sucrose) is widely used as a sweetener. Glucose, however, is sweeter-tasting and therefore required in smaller amounts. Glucose can be made from sucrose in one of the following ways: (i) Boiling a sucrose solution with dilute hydrochloric acid. (ii) Adding an enzyme (invertase) to a sucrose solution, kept at 35°C.

   (a) If glucose is prepared by treating sucrose with hydrochloric acid, what further treatment must be given to the glucose before it can be eaten?

   (b) What are the advantages of using an enzyme to produce glucose from sucrose?

2. Test five of the following foods for (a) reducing sugars and (b) non-reducing sugars: turnip, radish, parsnip, carrot, cabbage, peas, broad beans, washed baked beans. Tabulate your results.

4 Reducing sugars, starch and proteins in foods

TIME
Investigation: 30–40 minutes

In this investigation you will be using three different reagents to find out if reducing sugars, starch and proteins are present in different foodstuffs. The following reagents give a characteristic colour change in the presence of specific nutrients.

1 *Clinistix* contains an indicator which gives a blue colour with reducing sugars (glucose, fructose, maltose). The use of this reagent is described in Investigation 3.
2 *Iodine solution* produces a blue/black colour with starch. Place the food to be tested on the white tile and add a few drops of iodine solution from the dropping bottle.
3 *Albustix* contains an indicator which gives a green colour with protein. To test a food for protien, dip the yellow portion into water and crush the food to be tested on this part of the strip. Use a different reagent strip for each test.

You are provided with some foodstuffs and asked to find out if they contain reducing sugars, starch and protein. Bear in mind that each foodstuff may contains more than one of these nutrients. Therefore cut the foodstuffs into 2–3 smaller pieces before starting the investigation.

Investigation

Materials
- Six Clinistix reagent strips
- Five Albustix reagent strips
- Iodine solution
- 100 cm³ beaker of tap water
- Two portions of white bread
- Slice of banana
- Slice of cucumber
- Portion of cheddar cheese
- Two soaked raisins/sultanas
- White tile
- Scalpel

Method
1 Carry out the tests on the foodstuffs provided. Complete the tests with Clinistix and Albustix before testing for starch. When testing bread for starch, apply the test to only one of the portions. **Copy, and complete Table 3, below** (15)

2
(a) **Which foodstuff contains most reducing sugar?** (1)
(b) **Which foodstuff contains most protein?** (1)
(c) **Which foodstuff contains most starch?** (1)

Table 3 *Food tests for reducing sugars, protein and starch*

Test	Food	Colour change in Clinistix (Reducing sugars present/absent)	Colour change in Albustix (Protein present/COPPER absent) CAUSTIC SULPHATE SODA	Starch test (Starch present/absent)	P.H.
(a)	White bread	NIL	✓	✓✓✓	
(b)	Banana	TRACE ✓	30 MGS ✓	✓✓✓	6
(c)	Cucumber	TRACE ✓	TRACE ✓	✓	6
(d)	Cheese	NIL	30 MGS ✓	NIL	6
(e)	Raisin/sultana	·50 ✓✓	TRACE ✓	NIL	6
F.	POTOES	too MGS TRACE	✓✓	✓✓	6
G.	APPLE	✓	✓	✓	6

3 Chew the remaining portion of bread until it is thoroughly mixed with saliva. Put the chewed bread on one side of the white tile and leave it for a few minutes. Use a Clinistix reagent strip to test the chewed bread for reducing sugars.

(a) **What substance is present in the bread that was not there before?** *(1)*

(b) **How do you explain the appearance of this substance?** *(1)*

Taking it further

1 At each meal during the day, keep back a little of everything you will eat. Test each individual item of food in your diet for reducing sugars, starch and proteins. Copy Table 4, below, and enter your results. Is the carbohydrate/protein content of your diet balanced? What more do you need to know before it is possible to decide if you are eating a fully-balanced diet?

Table 4

Meal	Items of diet	Nutrients (✓ = present, X = absent)		
		Reducing sugars	Starch	Proteins
(a) Breakfast				
(b) Lunch				
(c) Tea/supper				

5 Estimation of vitamin C

TIME
Investigation: 40–60 minutes

Vitamins are an important part of the human diet. One of these vitamins, ascorbic acid or vitamin C, is contained in fruits and fresh vegetables. The recommended daily-intake of vitamin C for a human adult is 50–70 mg. A lower intake of the vitamin may prevent wounds from healing, slow the growth of bones, or weaken the walls of blood vessels. Vitamin C deficiency causes the disease called scurvy.

In this investigation you will use a blue dye called DCPIP (2, 6-dichlorophenol-indophenol) to find the amount of vitamin C contained in tablets and in grapefruit juice.

Investigation

Materials

- Vitamin C tablet, labelled A
- Vitamin C tablet (containing a different amount of the vitamin), labelled B

- Four DCPIP tablets
- 15 cm³ grapefruit juice
- 1 g sodium hydrogen carbonate (bicarbonate)
- 150 cm³ water in a beaker
- Two pestles and mortars (one exclusively for the DCPIP solution)
- Two 100 cm³ beakers (with a 50 cm³ graduation)
- 10 cm³ and 1 cm³ plastic syringe
- Boiling tube in a test-tube rack
- Litmus paper
- Test-tube holder
- Bunsen burner
- Forceps
- Wash-bottle, containing distilled water
- Glass beads
- Glass rod
- Knife
- Safety spectacles

Method

SAFETY PRECAUTIONS

Wear safety spectacles and a laboratory coat.

Part A

1. Crush tablet A in a mortar. Tip the crushed tablet into a 100 cm³ beaker. Use the wash-bottle to rinse out the mortar. Tip the washings into the beaker, then add water upto the 50 cm³ mark. Stir the vitamin C solution with a glass rod until the crushed tablet has dissolved.

2. Crush a DCPIP tablet in the clean mortar. Fill the 10 cm³ syringe with water, and add it to the crushed tablet. Grind the mixture until you have a clear blue solution, without any solid particles.

3. Draw some of the vitamin C solution into the small syringe, upto the 1 cm³ mark. Add this solution, drop by drop, to the blue dye in the mortar until the mixture changes from blue to red and then becomes colourless. (If necessary, stir the mixture with a glass rod, and refil the syringe if more than 1.0 cm³ vitamin C solution is required to decolorise the dye.)

4. The blue tablet you crushed contains the exact amount of dye made colourless by 1 mg vitamin C. Suppose you required 0.5 cm³ of the vitamin C solution to decolorise the blue solution. This would mean that 0.5 cm³ of the solution from the beaker contained 1 mg. It would also mean that the tablet contained $1.0 \times \dfrac{50}{0.5} = 100$ mg vitamin C.

Fig. 6 *Steps in finding out the amount of vitamin C in a tablet*

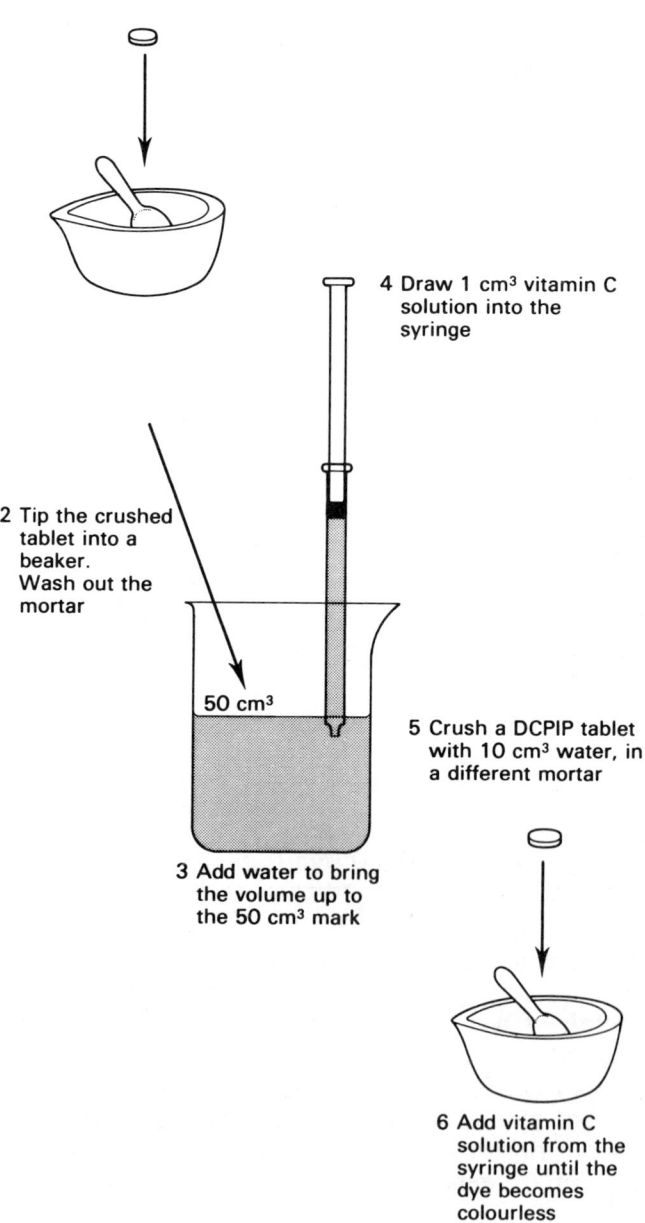

(a) **What was the volume (cm³) of vitamin C solution (A) you required to decolorise one crushed DCPIP tablet?** (2)

(b) **Calculate the amount of vitamin C (mg) in tablet A. Show how you arrive at your answer.** (3)

5 Take the vitamin C tablet labelled B and repeat instructions 1, 2, and 3.

(a) **What was the volume (cm³) of vitamin C solution (B) you required to decolorise a DCPIP tablet?** (2)

(b) **Calculate the amount of vitamin C in tablet B. Show how you arrive at your answer.** (3)

Part B

6 You have some grapefruit juice, which contains vitamin C. Repeat instructions 1, 2 and 3 of the investigation to find the volume of grapefruit juice that contains 1 mg vitamin C.

(a) **What is the volume (cm³) of grapefruit juice required to decolorise one DCPIP tablet?** (1)

(b) **Calculate the amount of vitamin C in 100 cm³ grapefruit juice. Show how you arrive at your answer.** (1)

7 Using forceps, dip a piece of litmus paper into the grapefruit juice.

(a) **What is the colour of the dipped paper?** (1)

(b) **What does this colour indicate about the chemical nature of grapefruit juice?** (1)

8 Pour 10 cm³ grapefruit juice into the boiling tube. Add 1–4 knifepoints of sodium hydrogen carbonate (bicarbonate), until the mixture turns the litmus paper blue.

(a) **What did you observe?** (1)

(b) **Suggest a reason for your observation.** (1)

Add a small, glass bead to the grapefruit juice in the boiling tube. Light the bunsen burner. Hold the tube in a test-tube holder, and heat the mixture until it boils. Cool the tube rapidly by running cold water from a tap on the outside of the glass. Wash out the mortars. Repeat instructions 1, 2 and 3 to find the amount of vitamin C in the boiled grapefruit juice.

(c) **What is the volume (cm³) of boiled grapefruit juice required to decolorise a DCPIP tablet?** (1)

(d) **Calculate the amount of vitamin C in 100 cm³ boiled grapefruit-juice. Show how you arrive at your answer.** (1)

(e) **What was the effect of the hot alkali on the vitamin C content of boiled grapefruit-juice?** (1)

9 Sodium hydrogen carbonate (bicarbonate) is sometimes added to green beans to preserve their colour during cooking. **What effect would this have on the vitamin C content of the beans?** (1)

Total 20 marks

Taking it further

1 In a hotel, segments of fresh grapefruit were prepared at breakfast and kept in the refrigerator until supper. A scientist, staying at the hotel, said that this would reduce the vitamin C content of the fruit. Design an experiment to find out if keeping prepared grapefruit in a refrigerator for 12 hours reduces its vitamin C content.
  (a) State your hypothesis.
  (b) How would you test your hypothesis?
  (c) What do you predict would happen if your hypothesis was correct?
  (d) Draw a table, with headings, for your results.
2 Which natural juice contains most vitamin C: orange, lemon, or apple? Does fruit juice, bought in cartons from supermarkets, contain the same amount of vitamin C as fresh fruit juice?

6 Sugar and salt in food

TIME
Investigation: 50–60 minutes

Many foods are flavoured by adding sugar or salt. In canned fruits, concentrated fruit-juices and squashes, glucose syrup (sugar) is one of the main ingredients. There is a lot of salt in cheese, chocolate, processed fish and canned-meat products. Eating too much sugar can lead to tooth decay and cause obesity. Too much salt in the diet can raise blood pressure. This, in turn, may make a person more likely to develop serious illnesses, including 'heart attack' and 'stroke'.

Two methods may be used for finding out the amount of sugar or salt in a solution. The first depends on differences in mass. A known volume of a solution of sugar or salt is heavier than the same volume of water. This means that the greater the concentration of sugar or salt in a solution, the greater its mass.

The second method depends on the rate at which water diffuses from a solution into absorbent paper. Water will diffuse more rapidly from a dilute solution of sugar or salt than from one which is concentrated.

This is an investigation of the amounts of glucose in different solutions.

Investigation

Materials

- 25 cm³ glucose solution
- 25 cm³ distilled water
- 11 cm³ orange squash
- 11 cm³ concentrated orange juice
- 10 cm³ plastic syringe
- Four flat-bottomed tubes
- Four strips of blotting paper
- Ruler, graduated in millimetres
- Top-pan balance
- Glass-marking pen
- Graph paper

Method

Part A

1 Copy Table 5. Label the flat-bottomed tubes from 1–4. Weigh the empty syringe and record its mass in the spaces 1–4 in row (d) of the table. Fill the syringe to the 10 cm³ mark with distilled water and weigh it. Record the mass of the syringe and water in space 1 (c). Subtract the mass of the syringe to give the mass of 10 cm³ water (space 1 (e)). Put 0.5 cm³ distilled water into tube 1. Empty the syringe.

2 Fill the syringe upto the 2.5 mark with glucose solution, and draw in distilled water to the 10 cm³ mark. Weigh the syringe and its contents, and record your results in space 2 (c) of the table. Calculate the mass of the solution and enter it in space 2 (e). Put 0.5 cm³ of the solution into tube 2.

Table 5

		Mixture No.			
		1	2	3	4
(a)	Volume glucose solution in syringe (cm³)	0.0	2.5	5.0	10.0
(b)	Volume water in syringe (cm³)	10.0	7.5	5.0	0.0
(c)	Mass of water or solution + syringe (g)				
(d)	Mass of empty syringe (g)				
(e)	Mass of water or solution (g)				
(f)	Concentration of glucose solution (g/100 cm³)	0.0	2.5	5.0	10.0

3 Fill the syringe to the 5.0 cm³ mark with glucose solution, then draw in distilled water to the 10 cm³ mark. Weigh the syringe and its contents. Complete column 3 of the table. Put 0.5 cm³ of the solution into tube 3. Empty the tube.

4 Fill the syringe to the 10 cm³ mark with glucose solution. Weigh the syringe and its contents. **Complete column 4 of the table, to fill spaces that remain.** Put 0.5 cm³ of the solution into tube 4. Empty the syringe. *(2)*

5 **Draw a graph of the mass of the solutions (vertical axis) against the concentrations of glucose (horizontal axis).** (The amount of glucose contained in each solution is shown in a row (*f*) of Table 5.) *(6)*

6 What would be the mass of solutions containing the following amounts of glucose?

(a) **7g/100 cm³** *(1)*

(b) **9g/100 cm³** *(1)*

7 Weigh 10 cm³ orange squash.

(a) **Use your graph to find how much glucose is present.** *(2)*

(b) **Weigh 10 cm³ concentrated orange juice. Use your graph to find how much glucose is present.** *(2)*

Part B

8 Put one strip of blotting paper into each of the flat-bottomed tubes, labelled 1–4. See Figure 7. After 15 minutes, take out the papers and measure how far the liquid has diffused up each strip.

(a) **Tabulate your results** *(3)*

(b) **What is the relationship between the distance water diffuses along the paper and the concentration of the solution?** *(1)*

9 **Comment on the two methods used to compare amounts of glucose in solution.** *(2)*

Total 20 marks

Taking it further

1 Fishes such as sardines, pilchards and salmon are canned in brine, a concentrated solution of salt. When cans are opened, only a few cm³ of brine are available for testing. Design an experiment to find out the amount of salt in the brine from a can of fish.
  (a) List the materials and apparatus for your experiment.
  (b) Write out your method, listing instructions in the order in which they should be carried out.
  (c) Draw a table, with headings, for your results.

Fig. 7 *Method of comparing concentrations of glucose in solution*

The rate at which the water front rises depends on the concentration of glucose in the solution.

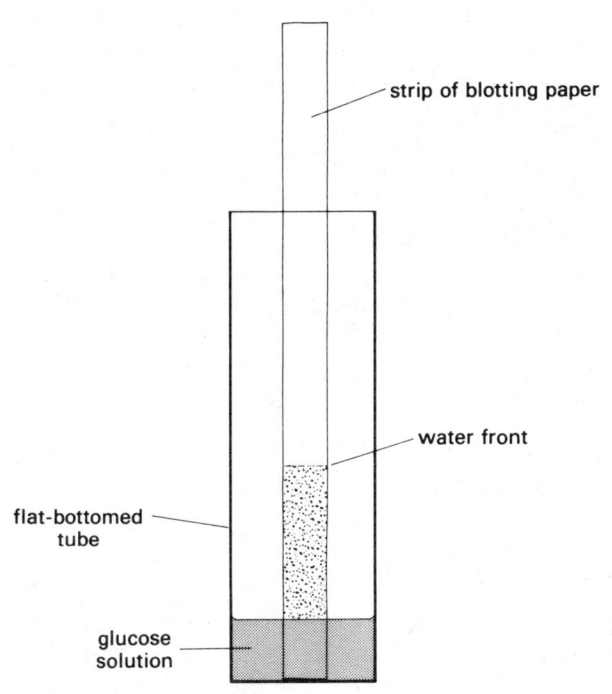

7 Spoilage of food by micro-organisms

TIME	
Preparation: 15–20 minutes	
Investigation: 30–40 minutes	

Method

SAFETY PRECAUTIONS
Wear safety spectacles

Micro-organisms are very small organisms, which include bacteria and some fungi. There may be thousands of spores of bacteria and fungi in every cubic metre of air. Each of these is capable of growing into a colony if it falls on a moist surface, where carbohydrates, fats, and proteins are present. If our food is left out in the air, bacteria and fungi can grow on it. Food spoiled by the growth of micro-organisms is unfit for human consumption. In addition to tasting unpleasant, such food may contain harmful chemical compounds, or carry organisms that can cause disease.

This investigation demonstrates the presence of bacterial and fungal spores in the air, and shows how they multiply on a suitable source of food. It also illustrates one way in which food may be preserved from attack by micro-organisms.

1. Remove the lid from the plate of agar, and expose it to the air for 5–10 minutes. Replace the lid and incubate the plate at 35 °C for 3–5 days.
2. Bruise one-half of the banana, peach, or orange with your finger or thumb; apply firm pressure but don't break the skin. Use the scalpel to cut small wedge-shaped pieces from the other half. Leave the fruit in a warm place, exposed to air, for 3–5 days.
3. Label the test tubes from 1–3. Cut three small cubes of potato, each about 1.5 cm^3. Put one cube of potato into each test tube. Add 5 cm^3 water to each tube. Hold tube 2 in a test-tube holder, and apply gentle heat from a bunsen burner until the water boils. Plug the neck of tube 2 with cotton wool and return it to the rack. Add the sodium chloride to tube 3. Allow the tubes to stand on the bench surface for 3–5 days. See Figure 8.

Preparation

Materials

- Banana, peach or orange
- Potato
- Plate of nutrient agar (or malt agar)
- Three test tubes in a rack
- 1 g sodium chloride
- 5 cm^3 plastic syringe
- Bunsen burner
- Incubator, maintained at 35°C
- Test-tube holder
- Ruler, graduated in millimetres
- Scalpel
- Cotton wool
- Glass-marking pen
- Safety spectacles

Fig. 8 *Preparing Investigation 7*

(i) Treatment of the fruit

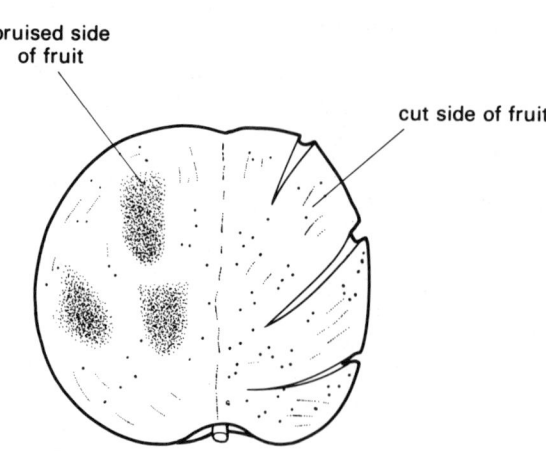

(ii) Setting up the test tubes

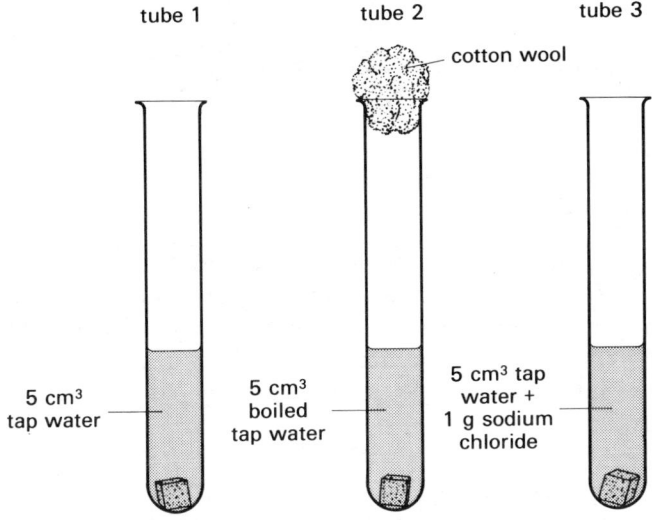

Investigation

Materials

Plate of agar, with colonies of micro-organisms. Banana, peach or orange, with bruises and cuts. Test tubes, containing cubes of potato

Method

1 Examine the plate of agar, without lifting the lid. You should be able to see two different types of colony. The small, round ones are bacteria. Larger colonies, some usually covered by a grey-brown powder, are fungi.

(a) **Make a labelled drawing to show the appearance of the agar plate. Label one bacterial and one fungal colony.** (2)

(b) **Describe how colonies of bacteria differ from colonies of fungi.** (2)

2 You have seen that bacteria and fungi can grow on an agar plate in only a few days. **Why should food not be exposed to air before it is eaten?** (2)

3 Examine the banana, peach, or orange to find if it has been attacked by micro-organisms.

(a) Describe what you see on the halves that were treated in the following ways:

   (i) **bruised**; (*1*)

   (ii) **cut.** (*1*)

(b) **What do you conclude from your results?** (2)

4 Examine the liquid in the three test tubes. **What do you observe in each tube, and how do you explain your observations?**

(a) **Tube 1** (2)

(b) **Tube 2** (2)

(c) **Tuble 3** (2)

5 A biologist believes a chemical compound in the skin of an orange kills bacteria and fungi. **How would you test this hypothesis?** (4)

Total 20 marks

Taking it further

1 Some cheeses are flavoured by bacteria and fungi. If you were given some blue-veined cheese, how would you find out if it contained living bacteria and fungi?
  (a) List the apparatus you would require.
  (b) Write out your method, listing instructions in the order that they should be carried out.
  (c) Draw a table, with headings, for your results.

2 How would you find out if temperature affects the rate at which micro-organisms spoil food?

3 How would you find out if mould fungi grew more rapidly on a moistened slice of wholemeal bread, or a slice of modern processed white bread? Remember, after moistening the bread, slices should be kept in an air-tight container, such as a translucent lunch box.
  (a) List the materials and apparatus for your experiment.
  (b) Write out your method, listing instructions in the order they should be carried out.
  (c) Draw a table, with headings, for your results.

8 The souring of milk

TIME
Preparation: 5 minutes per day for 7 days
Investigation: 40–50 minutes

Milk contains water, carbohydrates, fats, proteins and mineral salts. Milk sugar, or lactose, is a carbohydrate. When fresh milk is left to stand at room temperature for several days, bacteria change lactose into lactic acid. This has a sour taste. As a result, older samples of milk taste sourer than fresh milk.

In this investigation you will measure changes in the pH of samples of milk, as lactose is changed to lactic acid. A purple dye, resazurin, can be used to find out if living bacteria are in the milk. If large numbers of bacteria are present the dye is decolorised.

Preparation

Materials

- Seven 100 cm^3 beakers or paper cups
- Fresh milk (required daily for 7 days)
- Glass-marking pen

Milking cows by machine

Method

1 Number the beakers or paper cups from 1–7.
2 Each day, over a period of 7 days, pour 7.5 cm^3 fresh milk into one of the beakers or paper cups. Allow the containers to stand on a bench in the laboratory, exposed to air.

Investigation

Materials

- Seven samples of milk, numbered 1–7
- Narrow range pH 4.0–7.0 indicator paper
- 50 cm^3 resazurin solution
- Seven flat-bottomed tubes
- 5 cm^3 plastic syringe
- Forceps
- Ruler, graduated in millimetres
- Glass-marking pen
- Graph paper

Method

Part A

1 Smell, and attempt to pour, the milk samples numbered 1 and 7. **How do they differ?** (4)
2 Using forceps, dip a piece of pH paper into each sample of milk. Record the pH of each sample. **Tabulate your results.** (4)
3 **Express your results as a graph. Plot the age of the milk samples (days) along the horizontal (X) axis.** (4)
4 **Calculate the mean (average) pH of the samples of milk. Show how you arrive at your answer.** (2)

Part B

5 Number the flat-bottomed tubes from 1–7. Make two horizontal ink marks on each tube, one at 1.5 cm from the bottom, and the other at 3.0 cm. Pour milk from beaker of cup 1 into tube 1, until it reaches the 1.5 cm mark. Similarly, pour milk from the other

beakers or cups into tubes which have the same number.

6 Using the 5 cm³ syringe, add resazurin solution to each sample of milk until the 3.0 cm mark is reached. See Figure 9. Observe and record the colour of each mixture. **Tabulate your results.** (2)

Fig. 9 *Applying the resazurin test to a sample of milk*

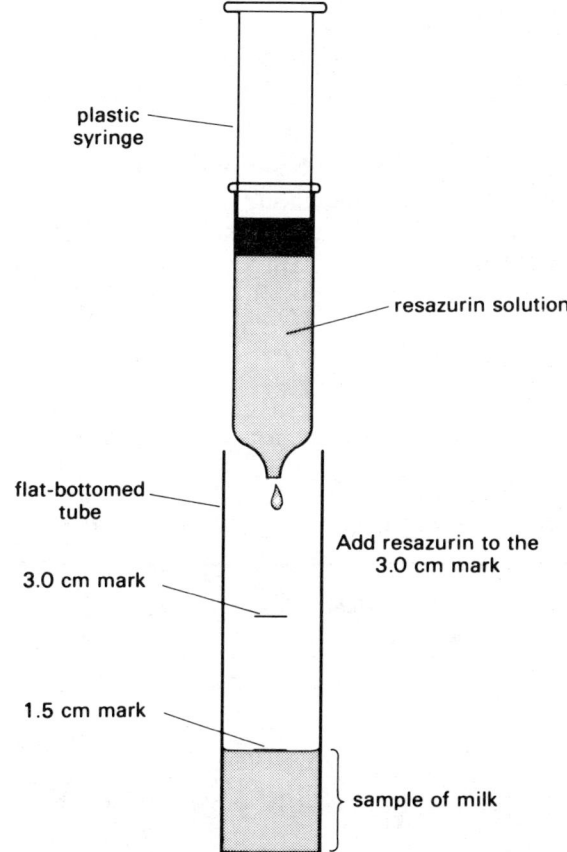

7 Allow the milk samples to which resazurin has been added to stand on the bench surface for 10–15 minutes. **What change did you observe?** (1)

8 Resazurin is used by dairies making daily deliveries of fresh milk. **Comment on the way you think the dye may be used.** (3)

Total 20 marks

Taking it further

1 How would you find out if long-life milk, available from supermarkets, stays fresh longer than the milk you have used?
   (a) State your hypothesis.
   (b) How would you test your hypothesis?
   (c) What do you predict would happen if your hypothesis was correct?
   (d) Draw a table, with headings, for your results.

Longlife milk, treated to preserve it from attack by bacteria

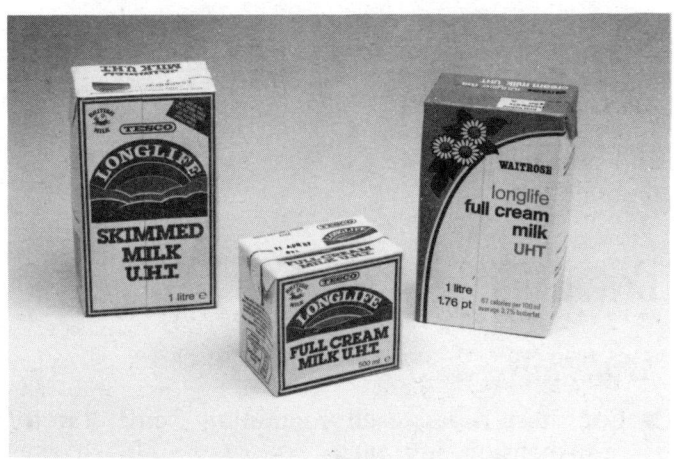

9 Environmental conditions affecting the digestion of starch

TIME
Investigation: 40–50 minutes

Starch is a carbohydrate which does not dissolve in water. It is digested by an enzyme called amylase. This enzyme breaks down molecules of starch into shorter units known as dextrins, then into single units of maltose which is a soluble sugar. Stages in the breakdown may be observed by adding iodine solution. A dark-blue/black compound is produced with starch. Dextrins form a light-blue/purple compound, while maltose doesn't change the brown colour of the iodine solution. When starch is digested the following colour change takes place:

starch + ⟶ dextrins + ⟶ maltose +
iodine   iodine   iodine
(dark-blue/black) (light-blue/purple) (brown)

This is an investigation to find out if temperature and other factors affect the rate at which starch is digested.

Investigation

Materials

- Four test tubes, each containing 5 cm³ starch suspension, in a rack
- 10 cm³ solution X
- 5 cm³ solution Y
- Iodine solution
- 250 cm³ beaker, containing ice
- 250 cm³ beaker
- Three 5 cm³ plastic syringes
- Four glass rods
- Four plastic straws
- White tile
- Stop-clock, watch with a second hand
- Bunsen burner
- Tripod and gauze
- Thermometer
- Paper towel
- Glass-marking pen

Method

1 Half-fill the beaker with water. Put the thermometer in the water. Set up the tripod and gauze. Light the bunsen burner. Heat water in the beaker to 40 °C, and keep it at this temperature. (Add cold water if the temperature rises above 40 °C, or apply more heat if it falls below.)

2 You have four test tubes, each containing 5 cm³ of starch suspension. Label the tubes A, B, C, and D. Put a plastic straw into each tube. Use a plastic syringe to add 5 cm³ solution X to tubes A and B. Mix well. Use a clean plastic syringe to add 5 cm³ solution Y to tube C, and 5 cm³ water to tube D. Mix well.

3 Mark the white tile as shown in Figure 10.

Fig. 10 *Marking and using the white tile*

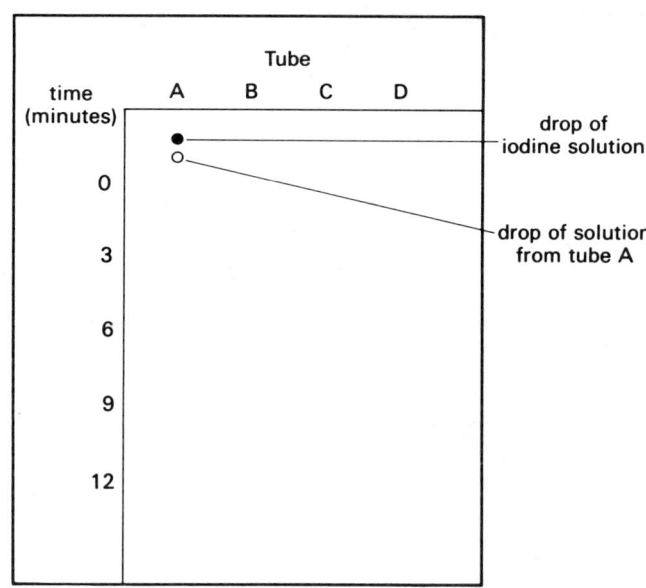

4 Put tube A into the beaker containing ice. Put tubes B, C and D into the beaker containing water at 40 °C. Start the stop-clock, or write down the time on your watch. Remove the straw from tube A. Touch the

bottom of the straw against the white tile, to put a drop of mixture A in the position shown in Figure 10. Put the straw back into tube A. Similarly, remove the other straws in turn and put one drop of mixtures B, C, and D along the same horizontal row of the tile. Return each straw to its tube.

5 Using the glass rod put a small drop of iodine solution next to each of drops A, B, C, and D on the tile. As quickly as possible after sampling, use the glass rod to mix the drop from tube A with iodine. Note any colour change. Rinse the rod and dry it on the paper towel.

Repeat the iodine test with drops from tubes B, C, and D, cleaning the rod after each test.

6 After 3, 6, 9, and 12 minutes, take out one drop from each mixture and test it with iodine solution. **Record your results in the form of a table, using the following symbols:**
   √√ **if the mixture is dark-blue/black**
   √ **if the mixture is light-blue/purple**
   0 **if the mixture remains brown** (5)

7 **In which of the tubes was starch digested?** (2)

8 Solution X (added to tubes A and B), and solution Y (added to tube C) were all prepared by dissolving the same substance in water. Solution Y was then given a further treatment.

(a) **What kind of substance was dissolved to give solutions X and Y?** (1)

(b) **What is the name of the substance in solutions X and Y?** (1)

(c) **How does solution Y differ from solution X?** (1)

(d) **What kind of treatment might have been given to solution Y?** (1)

9 **Explain any differences between:**

(a) **tubes A and B;** (2)

(b) **tubes B and C.** (2)

10 In this experiment tube D acted as a control.

(a) **What was the reason for setting up tube D?** (1)

(b) **Why was water added to the suspension of starch?** (1)

11 **When testing samples why was it necessary to:**

(a) **shake the tubes;** (1)

(b) **mix with a glass rod, and not a straw;** (1)

(c) **rinse and wipe the glass rod after each test?** (1)

Total 20 marks

Taking it further

1 The digestion of starch by amylase is affected by pH. Buffer tablets (pH 4.0, 5.0, 6.0, 7.0, 8.0, and 9.0) are available from manufacturers. A single tablet can be added to 10 cm^3 of a starch suspension to control pH. Design an experiment to find out the effects of pH on the rate of starch digestion by amylase.
   (a) List the materials and apparatus you would require.
   (b) Write out your method, listing instructions in the order in which they should be carried out.
   (c) Draw a table, with headings, for your results.

2 When starch is digested by amylase, a reducing sugar, maltose, is produced. Design an experiment to measure amounts of maltose in the solution at 5, 10, 15, and 20 minutes after a mixture of starch solution and amylase was set up.
   (a) List the materials and apparatus you would require.
   (b) Write out your method, listing instructions in the order they should be carried out.
   (c) Draw a table, with headings, for your results.

10 The effect of different temperatures on trypsin activity

TIME
Investigation: 60–80 minutes

Trypsin is a protein-digesting enzyme, produced by the pancreas. In this investigation the enzyme is used to digest casein, a white protein contained in powdered milk.

casein in milk $\xrightarrow{\text{trypsin}}$ amino acids
(white)   (colourless)

When all the white protein has been digested, the mixture becomes colourless. The aim of this investigation is to find out how four different temperatures affect the rate at which casein is digested.

Investigation

- 45 cm³ milk suspension
- 45 cm³ trypsin solution
- Four 250 cm³ glass beakers
- Two 5 cm³ plastic syringes
- Eight small test tubes
- Three tripods and gauzes
- Four thermometers
- Bunsen burner
- Stop-clock, or watch with a second hand
- Glass-marking pen

Method

Part A

1. Label the beakers from 1–4. Half-fill the beakers with water. Put a thermometer into each beaker. Light the bunsen burner. Put a gauze on top of each tripod. Leave beaker 1 on the bench surface. Place each of the other beakers on one of the tripods and gauzes. See Figure 11.

2. Measure the temperature of water in beaker 1. Put the burner under each of the beakers in turn to obtain the following water temperatures:
       beaker 2: room temperature + 10 °C
       beaker 3: room temperature + 20 °C
       beaker 4: room temperature + 50 °C.
   Try to maintain these temperatures by reapplying heat if the water baths begin to cool.

3. Label the test tubes from 1–8. Using a 5 cm³ syringe, put 5 cm³ trypsin solution into each tube. Put two tubes into each beaker, as follows:
       Beaker 1: tubes 1 and 2
       Beaker 2: tubes 3 and 4
       Beaker 3: tubes 5 and 6
       Beaker 4: tubes 7 and 8.

Fig. 11 *Setting up Investigation 10*

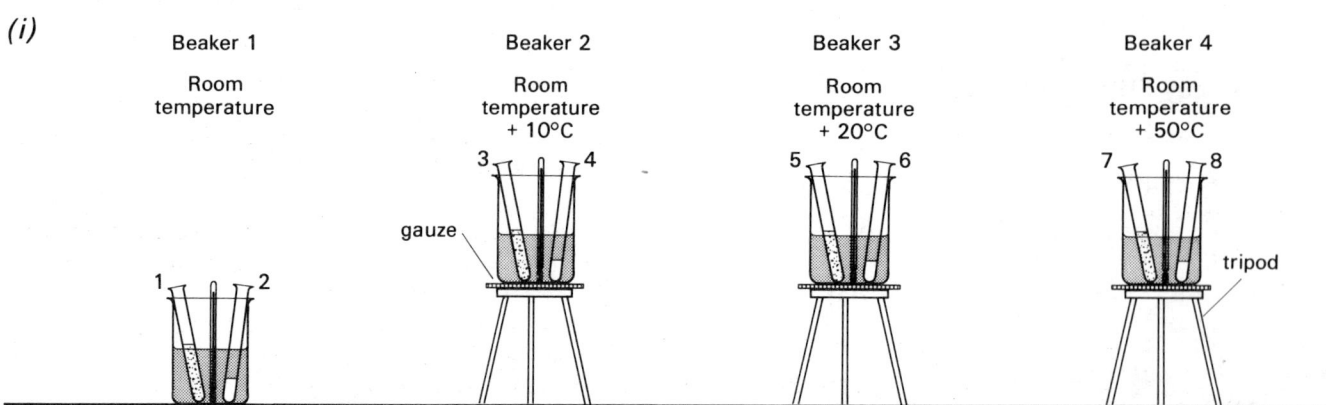

Fig. 11 (continued)

(ii) Enlarged view of beaker 1

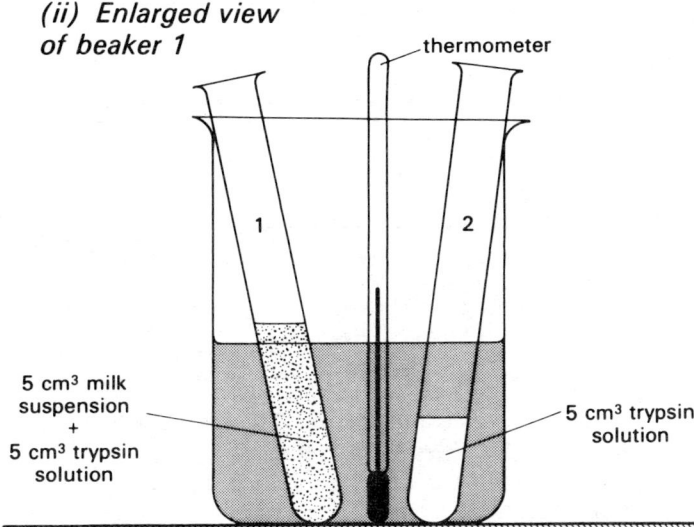

4 Using a clean syringe, put 5 cm³ milk suspension into tubes 1, 3, 5, and 7. Start the stop-clock, or write down the time.

5 Copy Table 6.

Enter the time taken for the milk/enzyme mixture in each tube to become colourless *(4)*

Table 6

Tube No.	Temperature of reaction (°C)	Time for completion of reaction (mins)
1	Room temperature	
3	Room temperature + 10	
5	Room temperature + 20	
7	Room temperature + 50	

6
(a) **At which temperature was the reaction completed most rapidly?** *(1)*

(b) **At which temperature was the reaction slowest, or not completed?** *(1)*

(c) The rate of many chemical reactions increases as temperature increases. **Why is this not true of reactions catalysed by enzymes?** *(2)*

7
(a) **Criticise the design of the experiment.** *(1)*

(b) **Suggest how the design of the experiment might be improved.** *(2)*

Part B

8 See Figure 11. In each beaker is a tube containing 5 cm³ trypsin. Add 5 cm³ milk suspension to each of these tubes. Transfer all of the tubes to the beaker at room temperatures +20 °C. Restart the stop-clock, or write down the time. Copy Table 7.

Enter the time taken for the milk/enzyme mixture in each tube to become colourless. *(4)*

Table 7

Tube No.	Temperature of incubation (°C)	Time for completion of reaction (mins)
2	Room temperature	
4	Room temperature + 10	
6	Room temperature + 20	
8	Room temperature + 50	

9 **What was the effect of keeping the enzyme at 50 °C above room temperature?** *(1)*

10 Biological washing powder contains a protein-digesting enzyme, which works best at a higher temperature than does trypsin. **How would you find out the optimum temperature for this protein-digesting enzyme in washing powder?** *(4)*

Total 20 marks

Taking it further

1 Biological washing powders contain protein-digesting and fat-digesting enzymes. Manufacturers recommend that clothes stained with gravy and egg yolk should be soaked in a solution of the powder, kept at 40 °C, for several hours. (Gravy stain is a protein; egg yolk is a fat.) How would you show that a biological washing powder washes clothes cleaner at 40 °C than at 70 °C?
(a) State your hypothesis.
(b) How would you test your hypothesis?
(c) What do you predict would happen if your hypothesis was correct?
(d) Draw a table, with heading, for your results.

11 Measuring enzyme activity by gas production

TIME
Investigation: 30–40 minutes

Catalase is an enzyme, present in all living cells, including dried yeast. Its function is to protect cells by breaking down a group of poisonous compounds called peroxides. For example, it will break down hydrogen peroxide into oxygen and water, both of which are harmless.

$$\underset{\text{hydrogen peroxide}}{H_2O_2} \xrightarrow{\text{catalase}} \underset{\text{oxygen + water}}{O_2 + 2H_2O}$$

One product of the reaction, oxygen, is a gas. Under normal conditions this gas would escape into the atmosphere, but in this investigation it is retained as a foam, formed from small bubbles. The oxygen is trapped as bubbles by mixing the enzyme, contained in the dried yeast, with washing-up liquid. A detergent in the washing-up liquid prevents the bubbles from escaping. As the reaction proceeds, a column of bubbles forms above the reagents. You will use the height of this column to show the rate of catalase activity. Remember that hydrogen peroxide is poisonous and can cause burns if it comes into contact with your skin.

Investigation

Materials

- Suspension of yeast cells, with added detergent
- 50 cm^3 hydrogen peroxide solution
- 50 cm^3 distilled water
- Iron filings (on a labelled sheet of paper)
- Granules of dried yeast (on a labelled sheet of paper)
- Seven test tubes, in a rack
- Spatula
- Ruler, graduated in millimetres
- Glass-marking pen
- Safety spectacles
- Plastic gloves
- Graph paper

Method

SAFETY PRECAUTIONS

Wear safety spectacles and plastic gloves. Wash off any spillages with plenty of water.

Part A

1 Put on safety spectacles and plastic gloves. Pour a little hydrogen peroxide solution into two test tubes. Using the spatula provided, add two or three iron filings to the hydrogen peroxide solution in one of the tubes.

(a) **What do you observe?** *(1)*

(b) **How do you account for your observation?** Using the spatula, add two or three granules of dried yeast to the hydrogen peroxide in the second tube. *(2)*

(c) **What do you observe?** *(1)*

(d) **How do you account for your observation?** *(2)*

Part B

2 Number the remaining test tubes from 1–5. Make five horizontal marks, 1 cm apart, from the bottom of the tube, as shown in Figure 12.

Return the tubes to the rack, and carefully pour hydrogen peroxide solution into tube 1 up to the level of the first mark. Similarly, pour hydrogen peroxide solution into tube 2 upto the level of the second mark; tube 3 upto the third; tube four up to the fourth; and tube 5 upto the fifth. Add distilled water to tubes 1–4 to bring up the volume of liquid in each tube to the same level as in tube 5. Using the syringe, add 1 cm^3 yeast suspension to each tube. As soon as a froth has formed, measure the height of the column of bubbles above the mixture in each tube.
Tabulate your results *(2)*

3 **Plot your results as a bar graph.** *(4)*

4 No control was used in this investigation. **Suggest a suitable control, giving reasons for your choice.** (2)

Part C

5 Design an experiment to find out the effect of adding different volumes of yeast suspension (up to $2.5\,cm^3$) to $2.0\,cm^3$ hydrogen peroxide solution, contained in each of five test tubes. **Write your instructions for carrying out the experiment, listing the different steps in order.** (6)

Total 20 marks

Taking it further

1 Orange and lemon juice both contain catalase. Design an experiment to find out if lemon juice contains more catalase than orange juice.
   (a) State your hypothesis.
   (b) How would you test your hypothesis?
   (c) What do you predict would happen if your hypothesis was correct?
   (d) Draw a table, with headings, for your results.

Fig. 12 *Setting up Investigation 11*

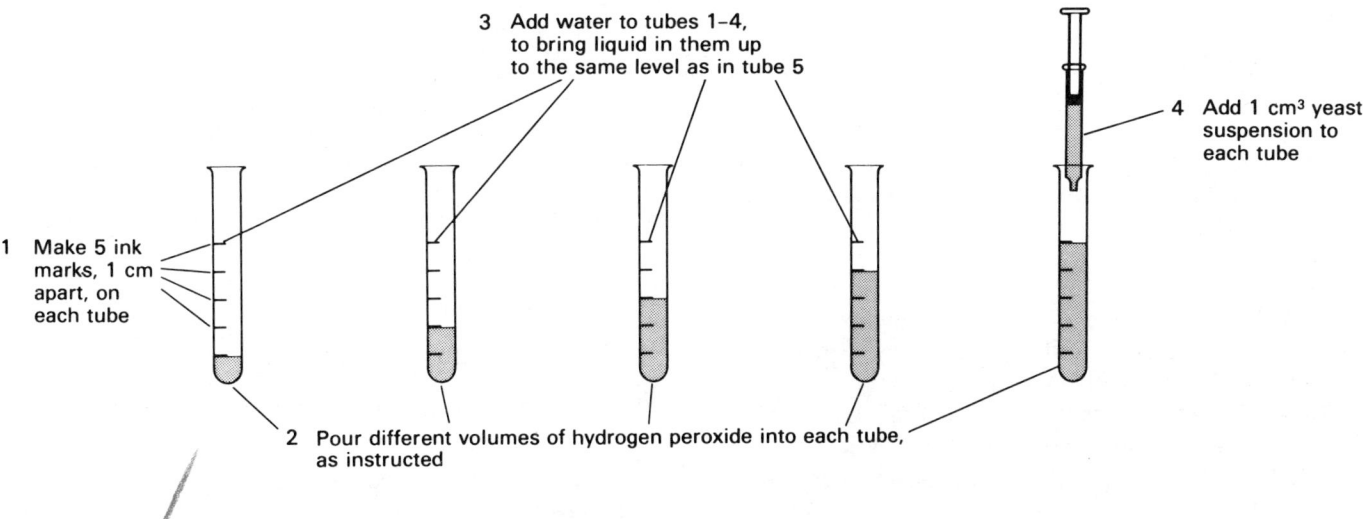

12 Using enzymes to extract fruit juice: an introduction to biotechnology

TIME
Preparation: 15–20 minutes
Investigation: 30–40 minutes

Enzymes are the catalysts of chemical reactions taking place in living organisms. Each enzyme is a protein that catalyses a particular reaction. Some enzymes cause large molecules to be broken down into smaller ones.

For thousands of years, humans have used microscopic living organisms, including bacteria and yeasts, to produce desirable products such as cheese, beer, and vinegar. Enzymes produced by the micro-organisms, however, were chiefly responsible for forming these useful substances. Today, the industrial use of micro-organisms is called biotechnology. Extracting and using enzymes for useful purposes has become a part of that technology.

You are going to investigate the possible uses of two enzymes, cellulase and pectinase, that break down plant cell-walls into soluble sugars:

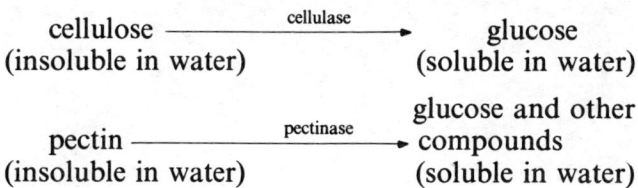

25

Preparation

Materials

- 400 cm³ apple purée, in a beaker
- Pectolytic enzyme powder (pectinase)
- Cellulase enzyme powder
- Four 100 cm³ measuring cyclinders
- Two 100 cm³ beakers
- Four filter funnels
- Four filter papers
- Teaspoon
- Tissue paper
- Glass-marking pen

Method

1. Place a filter funnel on top of each measuring cylinder, and number them from 1–4. Line each filter funnel with a folded filter paper.
2. Pour apple purée into one of the beakers up to the 100 cm³ mark. Tip this into filter funnel 1. Use the teaspoon to scrape out the beaker. Wipe the spoon on tissue paper.
3. Pour 100 cm³ apple purée into the beaker. Add one level teaspoon of cellulase powder and stir. Tip the mixture into filter funnel 2. Scrape out the beaker and wipe the spoon.
4. Pour 100 cm³ apple purée into the clean beaker. Add one level teaspoon pectolytic enzyme powder. Stir the mixture and pour it into filter funnel 3. Wipe the spoon.
5. To the remaining 100 cm³ apple purée, add one level teaspoon of cellulase powder and one of pectolytic enzyme powder. Stir the mixture and pour into filter funnel 4.
6. Stand the four measuring cylinders in a warm place for 12–24 hours.

Extracting the juice from apples

Investigation

Materials

- Four measuring cylinders, containing filtered apple juice

Method

1. **Make a table to show the volume of juice collected in each measuring cylinder.** *(4)*
2.
   (a) **Which measuring cylinder acts as the control?** *(1)*
   (b) **How does the control differ from other measuring cylinders?** *(1)*
3. **What is the percentage increase (or decrease) in the volume of juice extracted from the purée by adding the following?**
   (a) **cellulase** *(2)*
   (b) **pectinase** *(2)*
   (c) **cellulase and pectinase** *(2)*
   Show how you arrive at your answers.
4. **Why were the measuring cylinders stood in a warm place?** *(1)*
5. **What effect does pectinase have on extracted juice?** *(1)*
6. Juice extracted by adding these enzymes is often sweeter than natural juice. **Suggest a reason for this.** *(1)*
7. **Comment on the effect of the enzymes on juice extraction.** *(5)*

Total 20 marks

Taking it further

1. Cellulase and pectinase break down different parts of plant cell-walls. If you were provided with a dilute solution of each enzyme, how would you find out which one had most effect in breaking down the cell walls?
   (a) List the materials and apparatus for your experiment.
   (b) Write out your method, listing instructions in the order they should be carried out.
   (c) Draw a table, with headings, for your results.
2. Does boiling a fruit purée, before adding enzymes, increase the amount of juice that can be extracted?

13 Enzyme inhibition by drugs

TIME
Investigation: 40–50 minutes

Certain substances can reduce the activity of enzymes or stop them from working. These substances are called enzyme inhibitors. A number of drugs including alcohol (ethanol) and asprin (salicylic acid) have this effect. Good health depends on all our enzymes working properly. Therefore, if we regularly take some of these drugs, either in large amounts or over a long period of time, our health may be damaged.

The aim of this investigation is to show that two common drugs, alcohol and asprin, can affect the rate at which salivary amylase breaks down starch into maltose.

starch $\xrightarrow{\text{amylase}}$ maltose

(blue with iodine solution)   (no colour change with iodine solution)

It is important to realise, of course, that many other enzymes in our bodies may be affected by drugs in addition to acting as enzyme inhibitors, a number of drugs including alcohol, can stimulate excessive production of some enzymes.

Investigation

Materials

- Paper cup, containing 40 cm³ tap water
- 40 cm³ starch suspension
- 5 cm³ ethanol
- 5 cm³ salicylic acid (asprin) solution
- Iodine solution
- Eight flat-bottomed tubes
- Four 5 cm³ plastic syringes
- 1 cm³ plastic syringe
- Stop-clock, or watch with a second hand
- Glass-marking pen
- Graph paper

Method

1 Number the flat-bottomed tubes from 1–8, and arrange them in two rows on the bench, four per row. Using a 5 cm³ syringe, put 5.0 cm³ starch suspension into each tube. Fill a clean 5 cm³ syringe with ethanol and add 0.5 cm³ to tube 2, 1.0 cm³ to tube 3, and 2.0 cm³ to tube 4. Similarly, fill a third, clean 5 cm³ syringe with salicylic acid solution, and add 0.5 cm³ to tube 6, 1.0 cm³ to tube 7, and 2.0 cm³ to tube 8.

2 Take water from the paper cup into your mouth. Chew it over 40–50 times to get a sample of saliva rich in amylase, then spit it out into the cup. Using a clean 5 cm³ syringe, add 2.0 cm³ saliva to each tube, and gently rotate the tubes to mix their contents. Start the stop-clock, or write down the time.

3 After 10 minutes, use the 1 cm³ syringe to add 0.5 cm³ iodine solution to each tube. See Figure 13 on the next page. If colours are faint, add a further 0.5 cm³ iodine solution to each tube. Estimate the colour intensities of each tube using the scale in Table 8.

Table 8

Colour	Unit scale
Colourless	5
Brown, translucent	4
Brown, opaque	3
Brown-purple, translucent	2
Brown-purple, opaque	1
Blue	0

(a) **Tabulate your results for (i) ethanol and (ii) salicylic acid.** (4)

(b) **Plot your results as bar graphs.** (8)

4

(a) **What is the effect of increasing amounts of the drugs on amylase activity?** (2)

(b) **Which drug was the most effective inhibitor?** (1)

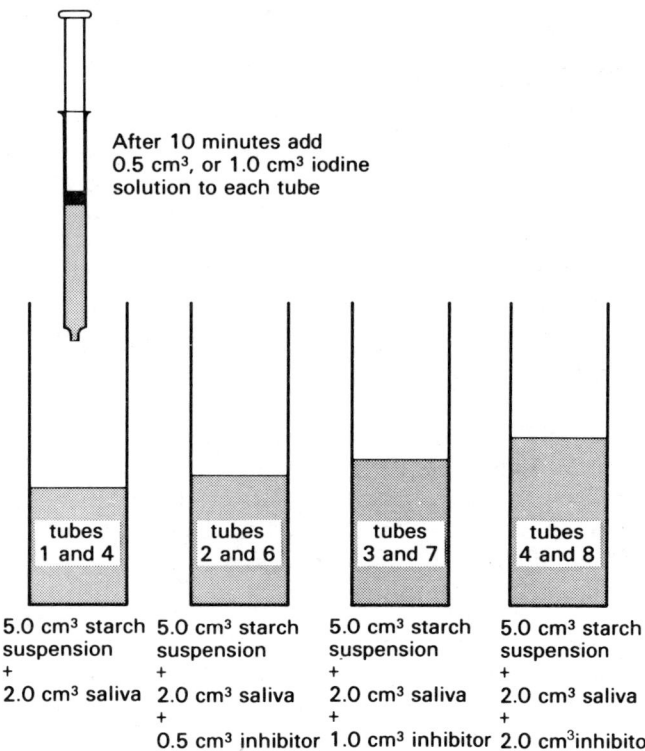

Fig. 13 *Setting up Investigation 13*

After 10 minutes add 0.5 cm³, or 1.0 cm³ iodine solution to each tube

tubes 1 and 4: 5.0 cm³ starch suspension + 2.0 cm³ saliva

tubes 2 and 6: 5.0 cm³ starch suspension + 2.0 cm³ saliva + 0.5 cm³ inhibitor

tubes 3 and 7: 5.0 cm³ starch suspension + 2.0 cm³ saliva + 1.0 cm³ inhibitor

tubes 4 and 8: 5.0 cm³ starch suspension + 2.0 cm³ saliva + 2.0 cm³ inhibitor

5 **Calculate the percentage of ethanol in each of the following mixtures:**

(a) **tube 2;** *(1)*

(b) **tube 4.** *(1)*

Show how you arrive at your answers.

6 **Criticise the design of the experiment, and suggest an improvement.** *(3)*

Total 20 marks

Taking it further

1 It has been observed that the presence of some metal ions in solution inhibits the activity of enzymes. A solution of copper (II) sulphate contains copper (Cu^{2+}) ions. How would you find out if copper ions inhibit the activity of amylase?
 (a) State your hypothesis.
 (b) How would you test your hypothesis?
 (c) What do you predict would happen if your hypothesis was correct?
 (d) Draw a table, with headings, for your results.

14 Sugar, bacteria and tooth decay

TIME
Investigation: 30–40 minutes

Eating large amounts of sugar can cause tooth decay. Our mouths contain several different species of bacteria. These bacteria produce enzymes that change sugar into acids.

sugars —[enzymes from bacteria]→ acids

If our teeth are not regularly cleaned, the acids can break down enamel and dentine, causing tooth decay.

This is an investigation of the cause of tooth decay. You will discover the reason for regularly cleaning your teeth. Your task is then to design an experiment to find out if people with the most bacteria in their mouths have the most dental fillings.

Investigation

Materials

- Two cotton buds, in a clean petri dish
- Toothpaste, on a white tile
- Universal indicator, in a dropping bottle
- Universal indicator colour-chart

Method

Part A

1 You have two cotton buds. Rub one across the top of your upper teeth, close to the gum. Hold the cotton bud over the open petri dish, and add two drops of universal indicator from the dropper. Put the used cotton bud into the petri dish. Rub the clean cotton bud up and down between your teeth.

Again, hold it over the petri dish, and add two drops of universal indicator. Put the cotton bud into the petri dish. Replace the lid of the petri dish.

(a) **What colour were your tooth scrapings after adding universal indicator?** *(1)*

(b) **What is the pH of your tooth scrapings?** *(1)*

(c) **What has caused the indicator to change colour?** *(1)*

(d) **Where does this substance come from?** *(1)*

(e) **How is the substance formed?** *(1)*

(f) **When your teeth are cleaned, how should they be brushed? Explain your answer.** *(2)*

2

(a) **Test and record the pH of the toothpaste on the white tile.** *(1)*

(b) **Comment on the pH of the toothpaste.** *(2)*

Part B

3

Figure 14 shows what happens if sucrose is added to saliva, and the mixture is kept at 37°C for 3–4 days. The pH falls, as bacteria in the saliva form acids from the sugar. Individuals differ in the rate at which these acids form, when 5 cm³ saliva and 5 cm³ sucrose solution are mixed and kept at 37°C.

Fig. 14 *Changes in the pH of a saliva/sucrose mixture*

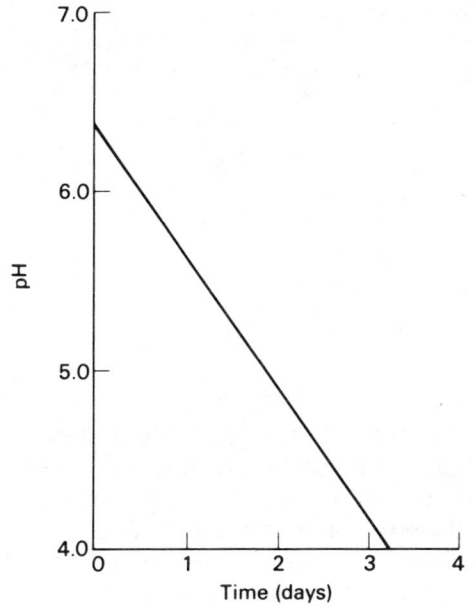

Design an experiment to find out if pupils with most acid-forming bacteria in their mouths have most dental fillings.
Your investigation is to be limited to ten pupils.

(a) **State your hypothesis.** *(2)*

(b) **How would you test your hypothesis?** *(4)*

(c) **What do you predict would happen if your hypothesis was correct.** *(2)*

(d) **Draw a graph to show how you would express the results you have predicted for a correct hypothesis.** *(2)*

Total 20 marks

Taking it further

1 Fluoride is added to some toothpastes to reduce tooth decay. The manufacturers claim it has reduced the number of fillings in children aged from 7–10 years. If you were given two brands of toothpaste, one containing fluoride, how would you test the manufacturers' claims on a group of 100 school children?
   (a) State your hypothesis.
   (b) How would you test your hypothesis?
   (c) What do you predict would happen if your hypothesis was correct?
   (d) Draw a table, with headings, for your results.
2 In some parts of the United Kingdom, fluorine is added to drinking water to reduce tooth decay. Argue the case for and against this practice. If you were in charge of your local water supply, what decision would you make? Give reasons for your answer.

Scanning electron micrograph (SEM) of colonies of bacteria growing on the surface of a tooth

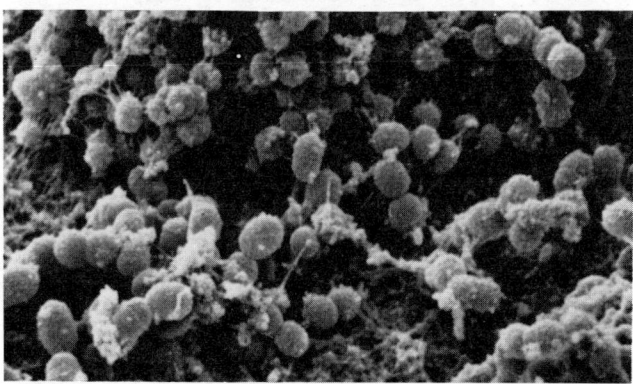

15 Demonstrating and making use of diffusion

TIME
Preparation: 15–20 minutes
Investigation: 30–40 minutes

The smallest particles of a gas are its molecules; those of a solution, its ions. These very small particles move continuously, from regions of high concentration to those where concentrations are lower. This process, called diffusion, can occur through gases, liquids, and some solids. Molecules diffuse through the air from food cooking in the oven. We can smell the food from a distance. As we get closer, the smell becomes more noticeable, because the number of molecules reaching our nose increases. Similarly, if ink is added to water, the ink diffuses outwards. The inky area gets bigger, but becomes less intensely coloured, as the ink spreads out.

This investigation shows ways in which the diffusion of molecules and ions can be observed. Stiff jellies, or gels, made from a substance called agar, are used. If the rates at which molecules and ions spread across a gel are compared, it is possible to find their approximate concentrations in solutions. You will measure the rates at which vitamin C diffuses from solutions of different strengths, then design an experiment to find out how much vitamin C is present in lemon juice.

Preparation

- Three Universal Indicator agar plates
- DCPIP agar plate
- 5 cm^3 vitamin C (ascorbic acid) solution
- 5 cm^3 distilled water
- Dilute hydrochloric acid, in a dropping bottle
- Dilute bench sodium hydroxide solution, in a dropping bottle
- Concentrated ammonia solution, and glass rod
- Three flat-bottomed tubes
- Three 1 cm^3 plastic syringes
- Cork-borer (No. 5, 6, or 7)
- Glass-marking pen
- Safety spectacles

Method

SAFETY PRECAUTIONS

Wear safety spectacles.

1 Put on your safety spectacles. Number the green Universal Indicator plates from 1–3. Turn plate 1 upside down. Lift the base of the dish so that the lid is exposed. Use a glass rod to put a single drop of ammonia on the lid. Return the base to its original position, with the gel above the drop of ammonia. Remove the lid from plate 2, and put a drop of ammonia on the surface of the gel, at the centre of the plate. See Figures 15(i) and 15(ii). Replace the lid.

Fig. 15 *Treatment of agar plates containing Universal Indicator*

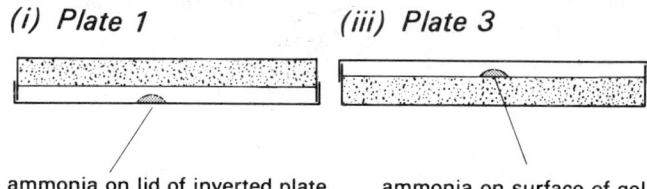

(i) Plate 1 — ammonia on lid of inverted plate

(iii) Plate 3 — ammonia on surface of gel

(ii) Plate 2

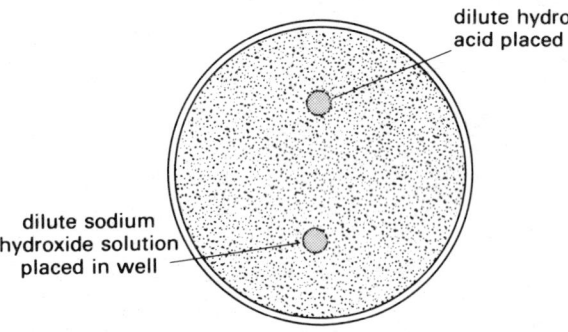

dilute hydrochloric acid placed in well

dilute sodium hydroxide solution placed in well

2 Use the cork-borer to cut two wells in plate 3, one on each side of the central position. (Tilt the borer gently to one side and lift when removing the unwanted piece of agar.) Put two drops of hydrochloric acid in one well, and two drops of sodium hydroxide solution in the other. Replace the lid,

and label the contents of each well. Allow the plate to stand on the bench surface for about 30 minutes.

3 The blue agar plate contains a dye called DCPIP. Cut four wells in this plate, spaced as shown in Figure 16, and number them. Use a plastic syringe to put two drops of the vitamin C solution into well 1. (This is a 4% solution of vitamin C.) Prepare 2% vitamin C solution by putting 0.5 cm³ vitamin C solution and 0.5 cm³ distilled water into a flat-bottomed tube. Gently shake the mixture, draw some into the syringe, and put two drops into well 2. Similarly, prepare a 1% vitamin C solution. Put 0.25 cm³ vitamin C solution and 0.75 cm³ water into a clean flat-bottomed tube. After mixing, draw some of the mixture into a syringe, and put two drops into well 3. Finally, for a 0.5% solution, put 0.125 cm³ vitamin C solution and 0.875 cm³ water into a clean flat-bottomed tube. Again, after mixing, draw a little of the mixture into the syringe and put two drops into well 4. Replace the lid of the dish and leave it on the bench surface for 30–40 minutes.

Fig. 16 *Appearance of the DCPIP agar plate after vitamin C has diffused from the wells. Radii of the areas cleared of dye are marked by arrows*

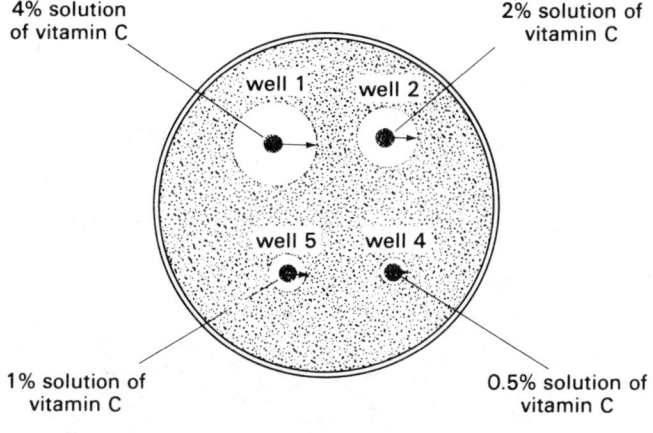

4% solution of vitamin C — well 1
2% solution of vitamin C — well 2
1% solution of vitamin C — well 5
0.5% solution of vitamin C — well 4

Investigation

Materials

- Three Universal Indicator agar plates
- DCPIP agar plate, containing vitamin C solutions

Method

Part A

1 Look at Universal Indicator plates 1 and 2. Do not remove their lids.

(a) **What colour is plate 1 immediately above the drop of ammonia?** *(1)*

(b) **What is the reason for the colour change?** *(1)*

(c) **How did the ammonia reach the agar?** *(1)*

(d) **What do you observe in plate 2?** *(1)*

(e) **How do you account for your observation?** *(1)*

2 Examine plate 3.
Describe the coloured circle surrounding

(a) **the well containing hydrochloric acid;** *(2)*

(b) **the well containing sodium hydroxide.** *(2)*

(c) **Which circle is larger?** *(1)*

(d) **Suggest a reason for any differences in size between the circles.** *(1)*

Part B

3 Look at the blue agar-plate. Measure the radii from the centre of each well to the edge of each circle cleared of dye. Use the formula πr^2 to calculate the total area of each colourless circle ($\pi = 3.14$). Copy Table 9. **Enter your results.** *(4)*

4 Lemon juice contains vitamin C. **If you were given some lemon juice, how would you use the plate of DCPIP agar, and the results from your table, to find out how much vitamin C it contained?** *(5)*

Total 20 marks

Table 9

Well No.	1	2	3	4
Concentration vit C (g/100 cm³)	4.0	1.0	0.5	
Radius of circle (cm)				
Area of circle (cm²)				

Taking it further

1 It is often said that lemon juice contains more vitamin C than orange juice. Design an experiment to test this hypothesis.
  (a) State your hypothesis.
  (b) How would you test your hypothesis?
  (c) What do you predict would happen if your hypothesis was correct?
  (d) Draw a table, with headings, for your results.

16 Diffusion of water through natural and artificial membranes

TIME
Preparation: 20–30 minutes
Investigation: 30–40 minutes

Cells are surrounded by membranes. These are sometimes said to resemble nets or sieves, with many pores. Small molecules, such as water, can pass through the pores. Larger molecules, such as glucose and starch, cannot. Therefore, membranes in living organisms are selectively permeable, or semi permeable, because they let through some molecules but not others.

All of the molecules in a liquid are moving. If a selectively permeable membrane separates a glucose solution from water, molecules constantly bombard the membrane from both sides. Any glucose molecules that strike the membrane rebound, and therefore do not leave the glucose solution. Water molecules, on the other hand, can pass through the pores in both directions. As these molecules are more concentrated on one side of the membrane than on the other, water diffuses into the solution of glucose. With the passage of time, the glucose solution becomes more dilute as its volume increases. See Figure 17.

This investigation demonstrates some properties of membranes, and the diffusion of water across them. One part of the investigation uses an artificial membrane of dialysis tubing, with pores that are larger than those found in natural membranes. The other uses the skins of dried fruits.

Fig. 17 *Diffusion of water 'molecules' through a selectively permeable membrane*

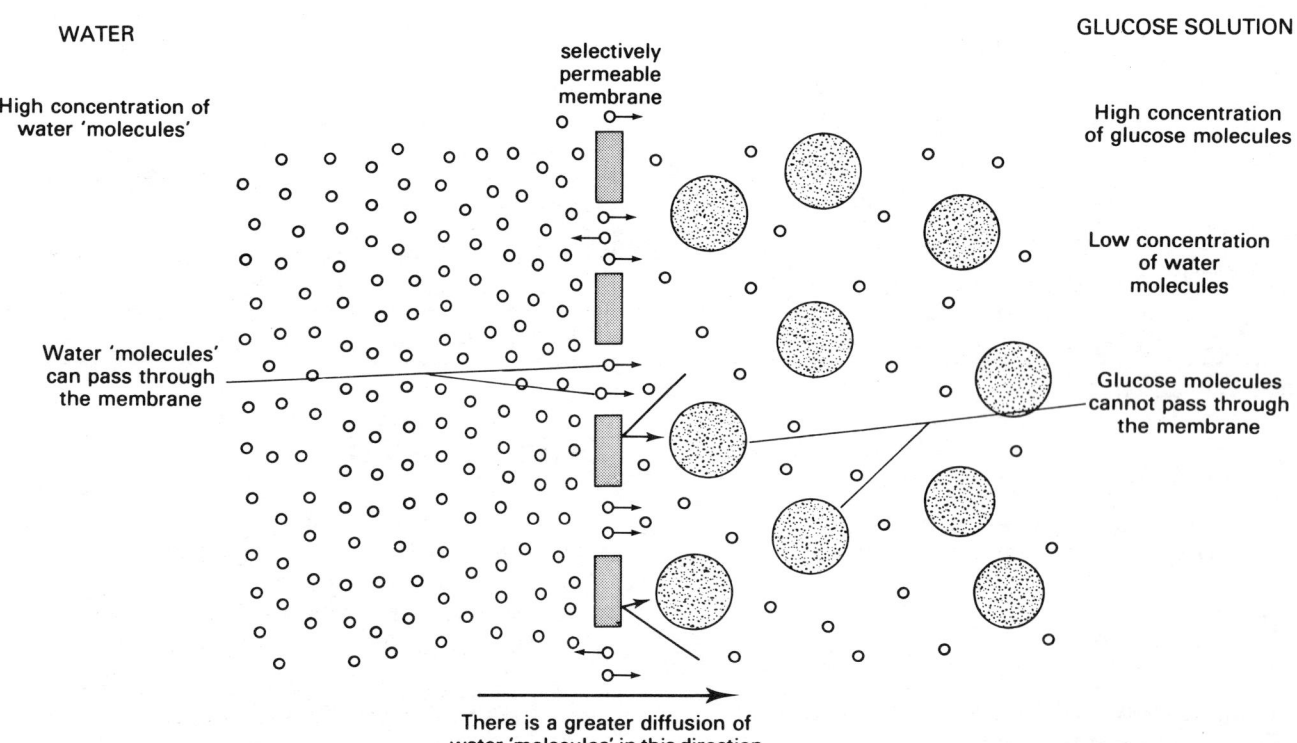

Preparation

Materials

- 25 g currants
- 25 g sultanas
- 25 g prunes
- 15 cm length of dialysis tubing
- 10 cm^3 starch/glucose mixture
- 20 cm^3 distilled water
- Boiling tube
- 250 cm^3 beaker
- Three 100 cm^3 beakers
- String
- Paper tissues
- Top-pan balance
- Glass-marking pen

Method

1 Half-fill the boiling tube with distilled water and put it in the 250 cm^3 beaker, as shown in Figure 18.

Fig. 18 *Arrangement of dialysis tubing in the boiling tube*

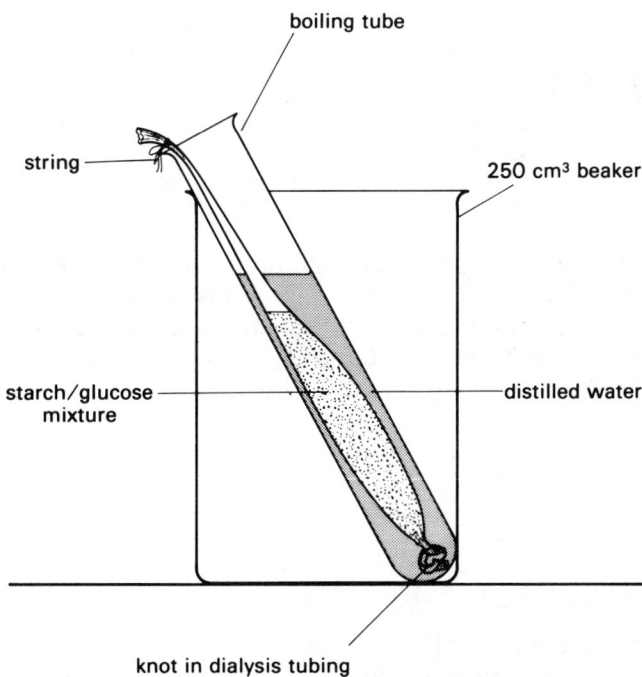

2 Tie a knot in one end of the dialysis tubing, and pour the starch/glucose mixture into it. Tie the open end of the dialysis tubing with string. Weigh the tubing, and write its mass on the outside of the beaker. Wash the outside of the tubing under a running tap. Put the dialysis tubing into the boiling tube. (Add more water to the boiling tube if the dialysis tubing is not surrounded by water.) Leave the apparatus for 24 hours.

3 Number the 100 cm^3 beakers from 1–3. Weigh out 25 g lots of currants, sultanas, and prunes. Put the dried fruit into beakers, as follows:

   Beaker 1: currants   Beaker 3: prunes
   Beaker 2: sultanas

Fill each beaker to within 1 cm of the rim with tap water. Allow the fruits to stand for 24 hours.

Investigation

Materials

- Apparatus containing dialysis tubing and starch/glucose mixture.
- Currants, sultanas, and prunes, soaked in water
- Clinistix reagent strip
- White tile
- Glass rod
- Paper tissues
- Top-pan balance

Method

Part A

1 **Why was the dialysis tubing washed under a running tap before it was put into the boiling tube?** *(1)*

2 Test the water outside the dialysis tubing for glucose and starch. Use the Clinistix reagent strip to test for glucose. Test for starch by removing one drop with the glass rod, and adding it to iodine solution on the white tile.
Record your results for:

(a) **glucose;** *(1)*

(b) **starch.** *(1)*

(c) **How do you explain your results?** *(2)*

3 Remove the dialysis tubing, dry it with a paper tissue, and weigh it.

(a) **Record both its original mass (from what you have written on the beaker) and its mass after standing in water for 24 hours.** *(1)*

(b) **What is the percentage increase, or loss, in mass? Show how you arrive at your answer.** *(2)*

(c) **How do you account for any change of mass?** *(2)*

(d) **What would have happened if more glucose had been put into the dialysis tubing?** *(1)*

Part B

4. Remove the currants from the water in beaker 1, dry them on tissue paper, and weigh them. Record their mass. Similarly, dry and weigh the sultanas and prunes. **Tabulate your results.** *(2)*

5 The original mass of each sample of dried fruit was 25 g. **Calculate the percentage increase in mass of each dried fruit. Show how you arrive at your answers.** *(3)*

6 **Which of the dried fruits has the greatest surface area through which water can diffuse?** *(1)*

7 **Which of the dried fruits is most likely to have the highest sugar content? Give a reason for your answer.** *(2)*

8 Drying fruits preserves them from being attacked by fungi and bacteria. **Why can't micro-organisms grow on dried fruit?** *(1)*

Total 20 marks

Taking it further

1 Fruits, such as peaches, are preserved by putting them into a concentrated solution of glucose, known as glucose syrup. Bacteria cannot grow in this concentrated solution, but they will grow if the glucose solution is diluted by adding water. Assume that you have some peaches in a molar solution of glucose (1.0 M). Design an experiment to find the lowest concentration of glucose in solution that will prevent bacteria growing.
  (a) List the apparatus and materials for your experiment.
  (b) Write out your method, listing instructions in the order they should be carried out.
  (c) Draw a table, with headings, for your results.

17 Water loss from leaves

TIME
Investigation: 60–90 minutes

Green land-plants lose water to the atmosphere. Most of this water loss occurs via the leaves. Minute pores in the epidermis of leaves, called stomata, are the principal sites from which water evaporates. Additional water loss may occur from cells in the epidermis. Although these cells are covered by a waxy cuticle, this serves only to restrict water loss, not to prevent it.

In many leaves there are differences in the rate of water loss from the upper and lower epidermis. This depends on several factors, such as the number, size and distribution of the stomata, or the thickness of the cuticle. The aim of this investigation is to compare rates of water loss from the upper and lower surfaces of the leaf, using two different methods.

Investigation

Materials

- Five leaves of cherry laurel
- Cobalt chloride paper
- Vaseline
- Cotton wool
- Adhesive tape

- Forceps
- Scissors
- Clock, or watch
- Top-pan balance
- Glass-marking pen

Method

Part A

1 Take four similar leaves of cherry laurel and number them. Use the cotton wool to apply a thin, continuous layer of vaseline over the surface of these leaves as follows:
Leaf 1: untreated;
Leaf 2: greased on upper surface;
Leaf 3: greased on lower surface;
Leaf 4: greased on both surfaces.
As soon as the vaseline has been applied, weigh each leaf and record its mass. Place the leaves in a dry, warm place. At intervals of 15 minutes, over a period of 45 minutes, reweigh the leaves and record their masses. **Tabulate your results.** *(5)*

(Set up part 6 of the investigation while you are waiting for results.)

2 **From your results calculate the total mass of water lost in 45 minutes from:**
(a) **Leaf 1;** (b) **Leaf 2;**
(c) **Leaf 3;** (d) **Leaf 4.**
Show how you arrive at your answers. *(2)*

3 **What percentage of its original mass has leaf 1 lost after 45 minutes? Show how you arrive at your answer.** *(2)*

4 **Was more water lost from the upper or lower surface of the leaf? Give a reason for your answer.** *(2)*

5
(a) **What was the effect of applying vaseline to the surfaces of the leaves?** *(1)*
(b) Both surfaces of leaf 4 were covered with vaseline.
From which part of this leaf can water continue to evaporate? *(1)*

Part B

6 Use forceps to put a piece of cobalt chloride paper on the upper surface of the fifth leaf, and cover it completely with a piece of adhesive tape, as shown in Figure 19. Similarly, put a second piece of cobalt chloride paper on the lower surface, and cover it with tape. Put a third piece of cobalt chloride paper on the bench surface, exposed to air. Record the time taken for each piece of cobalt chloride paper to change colour from blue to pink, as a result of absorbing water.

(a) **Tabulate your results.** *(2)*
(b) **What conclusions can be drawn from your results?** *(3)*

Fig. 19 *Applying cobalt chloride paper to the surface of a leaf*

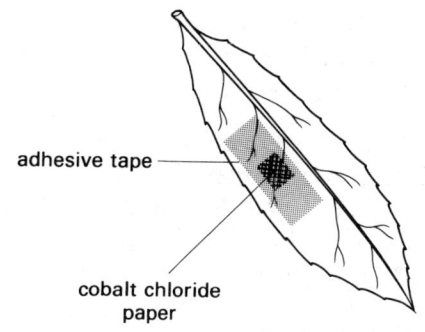

7 **Which of the methods used to measure water loss from leaves gave the best results? Give a reason for your answer.** *(2)*

Total 20 marks

Taking it further

1 It is claimed that beech leaves lose water faster than oak leaves. Design an experiment to test this hypothesis.
(a) State your hypothesis.
(b) How would you test your hypothesis?
(c) What do you predict would happen if your hypothesis was correct?
(d) Draw a table, with headings, for your results.

2 How would you measure the rate of water loss from a unit area (e.g. 10 cm² or 100 cm²) of leaf surface?

3 If a polythene bag is placed over a leafy shoot of a tree, and tied so that no water can escape, it is possible to find out how much water is lost from the shoot in a known period of time. Design an experiment to find out how much water is lost from the shoot in the periods (i) 6.00–12.00, and (ii) 13.00 p.m.–19.00.
(a) List the materials and apparatus for your experiment.
(b) Write out your method, listing instructions in the order they should be carried out.
(c) Draw a table, with headings, for your results.

18 Osmosis in living tissue

TIME
Preparation: 20–30 minutes
Investigation: 30–40 minutes

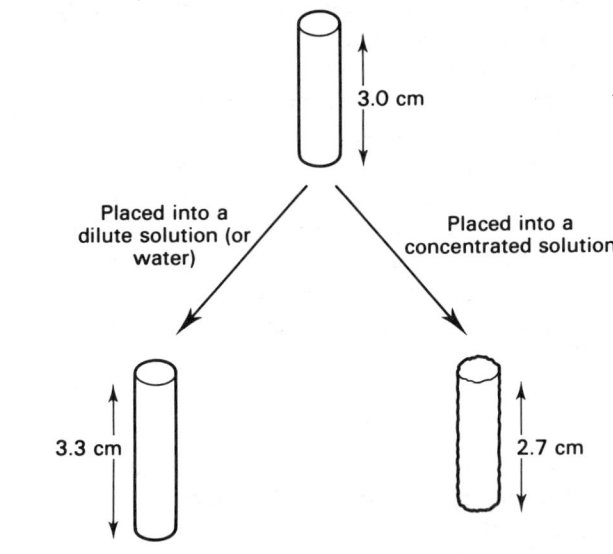

Fig. 20 *Changes in the appearance of rods of potato placed into dilute and concentrated solutions of sugar*

Cells contain a solution of substances, mostly sugars, dissolved in water. This is usually called a dilute solution of sugar, but could also be described as a solution in which water is concentrated. Water diffuses from regions where it is concentrated to regions where it is less concentrated. This movement of water is called osmosis. If cells are surrounded by a concentrated solution of sugar, water diffuses from the cell into the sugar solution. Alternatively, if cells are surrounded by a very weak solution, or by water, the movement of water is in the opposite direction.

Movement of water can be demonstrated by using solid rods of potato tissue, cut with a cork-borer. The effects of immersing these rods into water, or solutions of sugar, are shown in Figure 20.

The aim of this investigation is to measure and record changes in (i) the length of the rods and (ii) the volumes of the liquids in which the rods have stood.

Preparation

Materials

- Three or four large potato tubers
- 100 cm³ of each of the following:
    Distilled water (0.0 M)
    0.2 M glucose solution
    0.4 M glucose solution
    0.6 M glucose solution
    0.8 M glucose solution
    1.0 M glucose solution
- 100 cm³ measuring cylinder
- No. 13 cork-borer
- Six paper cups, or similar containers
- Ruler, graduated in millimetres
- Scalpel
- Glass-marking pen

Method

1. Number the cups from 1–6. Use the measuring cylinder to measure 90 cm³ distilled water, and pour it into cup 1. Pour 90 cm³ of each glucose solution into the other cups as follows:
    Cup 2: 0.2 M glucose solution
    Cup 3: 0.4 M glucose solution
    Cup 4: 0.6 M glucose solution
    Cup 5: 0.8 M glucose solution
    Cup 6: 1.0 M glucose solution

2. Use a No. 13 cork-borer (diameter 1.5 cm), or one of larger size, to cut solid rods of potato tissue. Trim each rod to 3.0 cm in length, and put four rods into each cup. Separate the rods so that they are covered by water or glucose solution. Allow the cups to stand on the bench surface for 24 hours.

Investigation

Materials

- Cups, containing rods of potato tissue
- 100 cm³ measuring cylinder
- Ruler, graduated in millimetres
- Graph paper

Method

1 Using the 100 cm³ measuring cylinder provided, measure the volume of distilled water, and of each solution of glucose, that now surrounds the rods of potato in each cup. **Tabulate your results.** *(4)*

2 **Draw a graph of your results. Label the vertical and horizontal axes, together with the 90 cm³ mark, as shown in Figure 21** *(4)*

Fig. 21 *Axes for the graph*
Distilled water has a molarity of 0.0

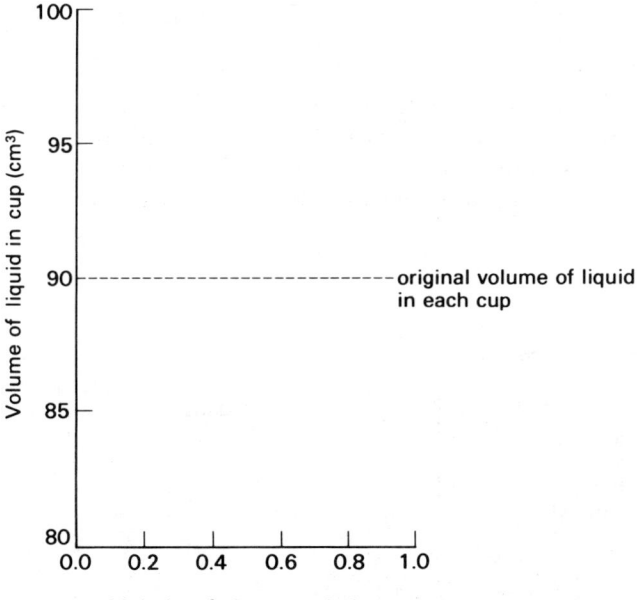

3 **Measure and record the length of one or more rods from each cup. Tabulate your results.** *(4)*

4 **Which of the cups contained solutions of glucose that were:**

(a) **less concentrated than the cell sap.** *(1)*
(b) **more concentrated than the cell sap.** *(2)*
(c) **of approximately the same concentration as the cell sap?** *(1)*

5 **In which direction does water diffuse when cells are surrounded by a solution of approximately the same concentration as the cell sap? Explain your answer.** *(2)*

6 **Suggest one way in which the design of the investigation could have been improved. Give a reason for the change you suggest.** *(2)*

Total 20 marks

Taking it further

1 Raw beetroot tastes sweeter than raw potato. Design an experiment to find out if the cell sap of beetroot contains more sugar than the cell sap of potato.
   (a) State your hypothesis.
   (b) How would you test your hypothesis?
   (c) What do you predict would happen if your hypothesis was correct?
   (d) Draw a table, with headings, for your result.

19 Using a simple potometer

TIME
Investigation: 30–40 minutes

Water is lost, in the form of vapour, from the leaves and stems of green land-plants. This continuous loss of water is called transpiration. Some biologists believe the process can be useful, especially during hot weather, as it cools the plant. Others believe it is more often harmful. Under hot, windy conditions plants may lose more water than they can take up. The result is wilting, a collapse of the leaves and stems as they become limp through loss of water. If wilting is not promptly relieved, either by reducing the rate of transpiration, or by supplying more water to the roots, plants may be permanently damaged, or die.

A simple potometer is an instrument used to measure the rate of water uptake by a shoot under different conditions. You are going to find out the effect of the leaves of a cut-shoot on its rate of water uptake.

Investigation

Materials

Cut-shoot of a large-leaved woody plant, bearing two leaves
100 cm^3 beaker, containing tap water
Simple potometer
Retort stand, boss, and clamp
Rubber tubing
Ruler, graduated in millimetres
Bench lamp, fitted with a 60 W bulb
Stop-clock, or watch with a second hand
Glass-marking pen
Graph paper

Method

1 Set up the potometer as shown in Figure 22, with the bench lamp at a distance of approximately 15 cm from the leafy shoot. Switch on the bench lamp. Apply gentle

Fig. 22 *Arrangement of the potometer and light source*

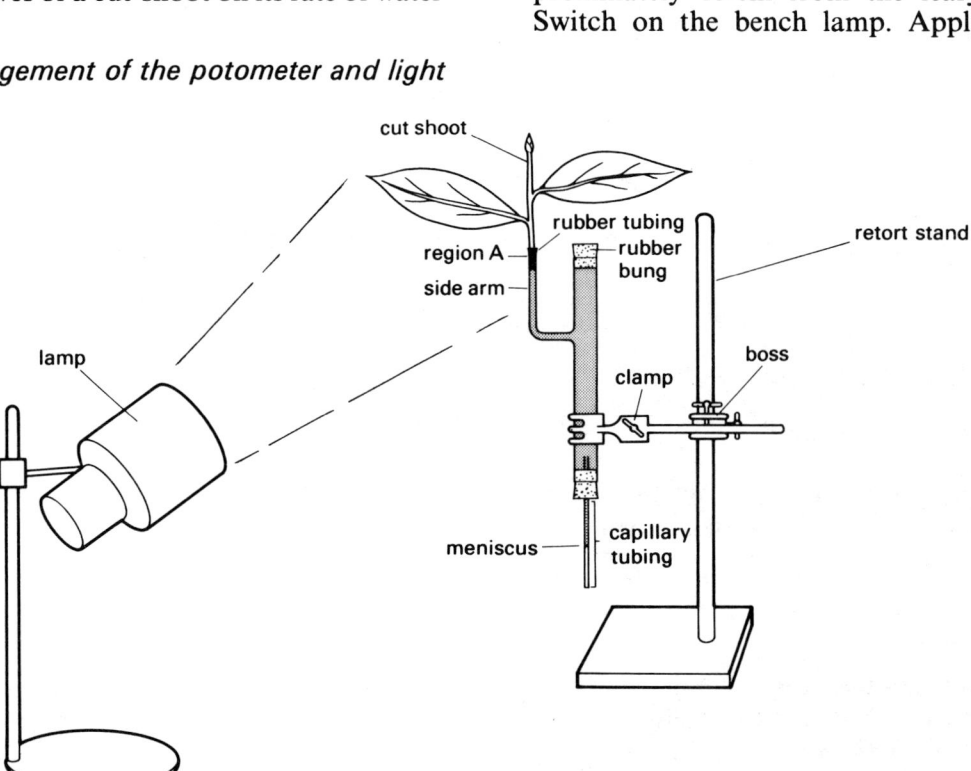

pressure to the rubber bung until the meniscus is pushed to the bottom of the capillary tube, and mark the position of the meniscus. Start the stop-clock and measure the distance travelled by the meniscus in two minutes. Record your result. Similarly, measure and record the distance travelled by the meniscus when one leaf is attached, and after all leaves have been removed. **Tabulate your results.** (3)

2 **Draw a graph of your results, plotting the distance travelled by the meniscus against the number of leaves removed.** (5)

3 When the lamp was switched on, what was the effect on the rate of water loss from the leaves of

(a) **light from the lamp** (1)

(b) **heat from the lamp?** (1)

4 The actual volume of water moving up the capillary tube may be calculated from the formula $\pi r^2 h$, where

$\pi = 3.14$
$r = $ radius of capillary tube
$h = $ distance (cm) travelled by meniscus in unit time

Assume the radius of the capillary tube to be 0.1 cm. From your results, calculate the actual volume of water passing through the capillary tubing over a period of 2 minutes, when both leaves were present. Show how you arrive at your answer. (2)

5

(a) **When all the leaves were removed, did the shoot continue to lose water?** (1)

(b) **Suggest a reason for the answer you have given.** (2)

6 Substances called antitraspirants are available from garden centres. When sprayed over plants, they provide a waterproof covering for leaves. **Explain why antitranspirants are used in each of the following situations:**

(a) **Reducing the fall of 'needles' from Christmas trees.** (2)

(b) **Assisting the rooting of cuttings.** (2)

7 **Suggest a further use for antitranspirants.** (1)

Total 20 marks

Taking it further

1 There are several different types of potometer. Some work better than others. Design an experiment to compare two different potometers.
   (a) List the materials and apparatus for your experiment.
   (b) Write out your method, listing instructions in the order they should be carried out.
   (c) Draw a table, with headings, for your results.
   (d) If you could afford to buy only one type of potometer, what would influence your choice?

2 Design an experiment to find out if a leafy shoot of oak takes up water more rapidly than a leafy shoot of beech.
   (a) State your hypothesis.
   (b) How would you test your hypothesis?
   (c) What do you predict would happen if your hypothesis was correct?
   (d) Draw a table, with headings, for your results.

20 Factors affecting the rate of photosynthesis

TIME
Investigation: 40–50 minutes
April–October

The rate of photosynthesis in all plants is affected by a number of external, or environmental, factors. The most important of these are:
1 light intensity;
2 amounts of carbon dioxide in the air (or water);
3 temperature.
In this investigation Canadian pondweed, a water plant, is used to demonstrate the effect of each of these factors. Cut-shoots of the plant, placed upside down in water, deliver bubbles of gas from their cut stems. The rate at which bubbles are produced indicates the rate at which photosynthesis is taken place.

Investigation

Materials

- Two cut-shoots of Canadian pondweed
- 5 cm³ sodium hydrogen carbonate (bicarbonate) solution
- 1 cm³ beaker
- 1 cm³ plastic syringe
- Two glass rods
- Cotton
- Thermometer
- Tripod and gauze
- Bunsen burner
- Scissors
- Stop–clock, or watch with a second hand
- Graph paper

Fig. 23 *Apparatus for investigating the effect of temperature on the rate of photosynthesis*

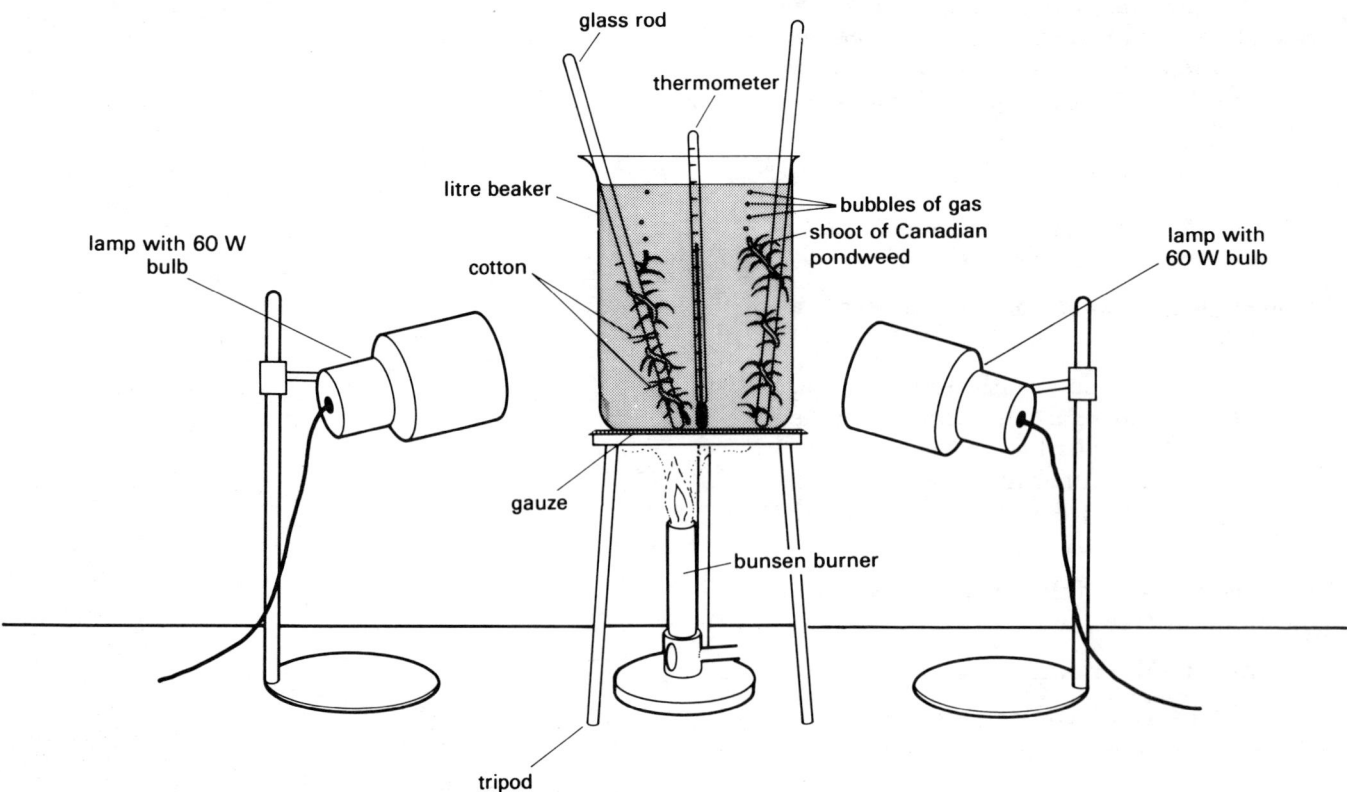

Method

1. Fill the beaker with tap water to within 2–3 cm of the rim. Take the glass rods and wind a piece of Canadian pondweed around one end of each rod, tying it in position with short lengths of cotton. Put the two rods into the beaker, with the shoots upside down, as shown in Figure 23. Put the thermometer into the beaker.

2. Gently lift the beaker on to the tripod and gauze. Switch on the bench lamps, and arrange them on opposite sides of the beaker, at a distance of 5 cm from it. Look at the cut ends of the stems, then choose one that is delivering bubbles at a rate of 15–35 per minute. (It may be helpful to cut off a few millimetres of stem from one of the shoots, using the scissors. This often changes the rate at which bubbles are delivered.)

3. Measure and record the number of bubbles delivered per minute when the bench lamps are at a distance of 5, 10, 15, 20, and 30 cm from the beaker. **Tabulate your results**. (2)

4. **Plot your results as a graph**. (4)

5. With the bench lamps placed at 10 cm from the beaker, record the number of bubbles delivered per minute. Add 0.25 cm^3 sodium hydrogen carbonate (bicarbonate) solution from the syringe, stirring with the thermometer. Record the number of bubbles delivered per minute. Similarly, record the number of bubbles produced per minute after making further additions of 0.5 cm^3, 1.0 cm^3 and 2.0 cm^3 hydrogen carbonate (bicarbonate) solution. **Tabulate your results**. (2)

6. **Plot your results as a graph.** (4)

7. Light the bunsen burner. Place it beneath the tripod and gauze. Count the number of bubbles delivered per minute at 20, 30, 40, 50, and 60 °C. **Tabulate your results**. (2)

8. **Plot your results as a graph** (4)

9. **Comment on the effects of temperature on the rate of photosynthesis.** (2)

Total 20 marks

Taking it further

1. The temperature at which photosynthesis takes place most rapidly is the optimum temperature. Design an experiment to find the optimum temperature for photosynthesis in Canadian pondweed.
   (a) List the materials and apparatus for your experiment.
   (b) Write out your method, listing instructions in the order they should be carried out.
   (c) Draw a table, with headings, for your result.

2. Design an experiment to compare the rates of photosynthesis in red, blue, and yellow light. You can use coloured sheets of plastic, stuck to the outside of beakers.
   (a) List the materials and apparatus for your experiment.
   (b) Write out your method, listing instructions in the order they should be carried out.
   (c) Draw a table, with headings, for your results.

Bubbles of oxygen-rich gas coming from the leaves and stem of Canadian pondweed (Elodea canadensis)

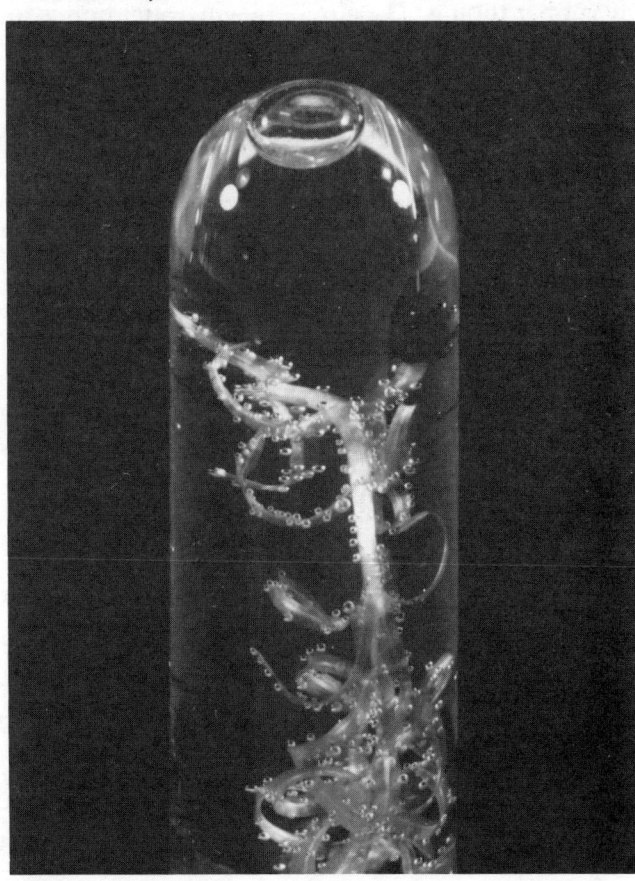

21 Gas production by Canadian pondweed during photosynthesis

TIME
Investigation: 40–50 minutes
April–October

Canadian pondweed, a submerged water plant commonly found in garden ponds and aquaria, produces bubbles of gas in bright light. Most of this gas is oxygen, formed in the leaves as a result of photosynthesis. The faster the rate of photosynthesis, the faster the rate at which bubbles are released. If these bubbles are trapped in the upturned barrel of a syringe, it is possible to measure the amount of gas produced in a known period of time, such as 5 or 10 minutes. As bubbles collect at the top of the barrel, an equivalent amount of liquid is displaced into a capillary tube, connected to the nozzle of the syringe by rubber tubing. The aim of this investigation is to find the effect of light intensity on the rate of photosynthesis.

Investigation

Materials

- Living leafy shoots of Canadian pondweed
- 25 cm^3 sodium hydrogen carbonate (bicarbonate) solution
- 20 cm^3 plastic syringe
- 100 cm^3 beaker
- 30 cm length of 0.1 mm diameter capillary tubing
- Rubber tubing (to fit over capillary and nozzle of syringe)
- Boss, clamp, and retort stand
- Bench lamp, fitted with a 100 W bulb
- Scalpel
- Ruler, graduated in millimetres
- Glass-marking pen

Method

1 Cut three leafy shoots of Canadian pondweed, each about 5 cm in length. Put these into the barrel of the 20 cm^3 syringe, with the cut end facing away from the nozzle, as shown in Figure 24. Clamp the syringe at 35–40 cm above the bench surface, nozzle downwards.

2 Use a piece of rubber tubing to attach the capillary tube to the nozzle end of the syringe.

3 Put the beaker beneath the capillary tube. Remove the plunger from the syringe. Put your thumb or finger over the open end of the capillary tube and pour sodium hydrogen carbonate (bicarbonate) solution into the syringe until it is full. Replace the plunger and gently apply pressure until the solution in the syringe reaches the 20 cm^3 mark.

Fig. 24 *Apparatus for measuring the volume of oxygen produced by Canadian pondweed during photosynthesis*

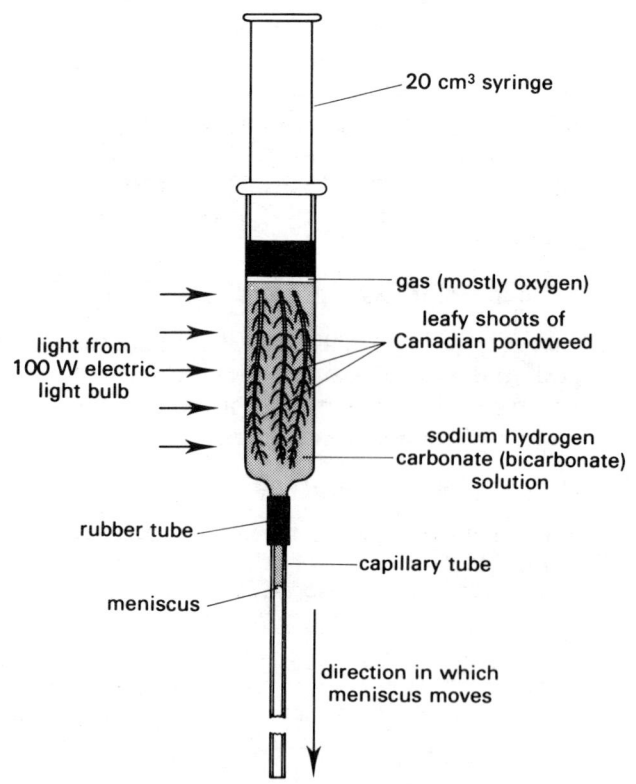

4 Gently raise the plunger to draw the meniscus near to the top of the capillary.

5 Switch on the bench lamp and put it at 20 cm from the syringe. Measure and record the distance travelled by the meniscus in 5 minutes.

6 Similarly, with the bench lamp at distances of 15, 10, and 5 cm from the syringe, measure and record the distance travelled by the meniscus in 5 minutes, when the lamp is in each position.

7 **Tabulate your results.** (4)

8

(a) **Sketch a graph to show the relationship between light intensity and the rate of photosynthesis.** (4)

(b) Why is it difficult to draw a graph from figures in your table? (2)

9 Suggest reasons for each of the following, when setting up the investigation:

(a) **using three cut-shoots of Canadian pondweed;** (2)

(b) **putting the shoots into sodium hydrogen carbonate (bicarbonate) solution;** (2)

(c) **illuminating the shoots with a 100 W electric light bulb.** (2)

10 A biologist has 20, 40 and 60 W electric light bulbs. **Design an experiment to find out the effect of light intensity on the rate of photosynthesis in Canadian pondweed. Write out your method, listing instructions in the order they should be carried out.** (4)

Total 20 marks

the light for 6 hours. After exposure to light, the same cork-borer is used to cut 100 more discs from the same leaves. These are weighed, and any change in mass calculated.

In a carefully controlled experiment with potted polyanthus plants, the results shown in Table 10 were obtained.

Table 10

Treatment of leaves	Mass of 100 leaf discs (g)
Before exposure to light	13.36
After exposure to light	14.41

(a) Calculate the total increase in mass.
(b) What was the mean (average) increase per hour?
(c) Do you think these results are very accurate? Give one or more reasons for your answer.

....

Taking it further
....

1 If a cut-shoot of Canadian pondweed produces 0.06 cm³ gas in 10 minutes, how much gas does it produce in (i) one hour, and (ii) 12 hours?

2 The rate of photosynthesis in potted land-plants can be found by cutting discs from leaves before and after exposure to light. A cork-borer is used to cut 100 discs from one-half of destarched leaves, kept in darkness for 12 hours. These are weighed and their mass recorded. The plant is then put in

22 Visits by bees to flowers

TIME
Preparation: 15–20 minutes
Observation: 2 days
Investigation: 30–40 minutes
April–September

Bees are useful insects: they provide us with honey and beeswax, but, much more importantly, they pollinate flowers. Crops such as fruits and many vegetables would fail if bees didn't pollinate their flowers, allowing fertilisation to take place. Many of the flowers in gardens are pollinated by bees. The bee visits flowers to feed on nectar, a sugary solution produced inside the ring of petals, and to collect pollen from the stamens. Plants that are pollinated in this way have to attract bees to their flowers. Additionally, the flowers must support a feeding bee, and direct it, by means of 'honey guides' to the source of nectar.

This is an investigation of the conditions that allow bees to visit flowers. It may be carried out as an individual project. Results may be obtained during the summer term, or holiday period, when plants are in flower, and bees are active.

Preparation

Materials

- Graph paper
- Cardboard
- Glue

Method

1 Cut out eight pieces of graph paper, each 10 × 15 cm, and stick each one onto a piece of stiff cardboard, to provide backing and support. Mark the axes of each piece of graph paper, as shown in Figure 25. Look carefully at Figure 25 and ask your teacher if you don't understand how the record sheets are filled in.

2 Read an account of flight and direction-finding by bees.

Fig. 25 *Record sheet of the number of bees visiting a patch of antirrhinums*

Seven bees, including one humble bee, visited the patch during the period of observation. A total of 14 flowers were visited.

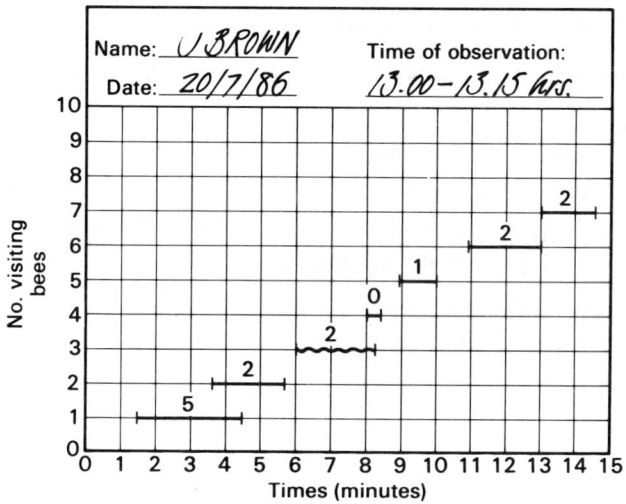

Observation

Materials

- Eight record sheets
- Stop-watch, or watch

Method

1 Select a small patch of a named flower, such as antirrhinums or petunias from a flower bed, strawberries or beans from a garden, or any patch of flowers growing in the wild.

2 On a sunny, calm day take up a position where you can observe all of the flowers in the patch. Make a record of visits by bees over four 15 minute periods: 10.00–10.15; 13.00–13.15; 16.00–16.15; 19.00–19.15 hrs. Use a straight line to record visits by hive bees, and a wavy line to record visits by humblebees. Draw lines of different lengths to show how long each bee spent at the patch. Put numbers above each line to record the number of individual flowers visited.

3 Return to the same patch on a dull, calm day, and make further records of visits by bees at the same times.

Investigation

Materials

- Eight completed record sheets
- Graph paper

Method

1 **Plot graphs from your records to show how the total number of bees visiting the flowers was affected by the time of day on:**

(a) **the sunny, calm day;** *(4)*

(b) **the dull, calm day.** *(4)*

2

(a) **Which type of bee was the most frequent visitor?** *(1)*

(b) **On which day did most bees visit the flowers?** *(1)*

(c) **Suggest reasons that could account for any differences in the total number of visits on the sunny and dull days** *(2)*

(d) **On the sunny day, at which time did most bees visit the flowers?** *(1)*

(e) **Suggest a reason for most bees visiting the flowers at this time** *(1)*

3 Suppose you made records on a sunny, windy day. **How might a strong wind affect your results?** *(2)*

4 **Hand the record sheets to your teacher for marking.** *(4)*

Total 20 marks

Taking it further

1 At what time of the year do bees make most visits to flowers? In addition to bees, what other insects act as pollinators?

2 You may prefer to observe visits made by garden birds to a bird table. Set up a bird table around your house, where it can be seen from behind a window. Use a mixture of breadcrumbs, crushed peanuts, and kitchen scraps as food. Put some food on the table, and scatter some on the ground.

(a) Use an illustrated guide to identify birds that eat the food.

(b) Which species feed (i) at the table and (ii) on the ground? Do any species feed in both positions?

(c) Do the same species come in summer and winter?

(d) Do more birds come in summer than winter?

(e) Find a method of keeping records to show (i) the mean (average) number of birds feeding together, and (ii) the length of time spent by each bird at the table.

Hive bee visiting a flower

23 Floral structure of tulip and polyanthus

TIME
Investigation: 30–40 minutes
March–May

Flowering plants belong to two major groups, the dicotyledons and monocotyledons. If you examine the flowers of a plant, it is possible to tell the group to which it belongs. Flowers of dicotyledons have distinct sepals, petals, stamens, and carpels. The different parts of these flowers are usually present in multiples of 4 or 5. Flowers of monocotyledons do not have distinct sepals and petals. Instead, there are one or two rings of petal-like perianth segments. The perianth segments, stamens, and carpels of these flowers are in multiples of 3.

The aim of this investigation is to compare and contrast the flower of a dicotyledon with that of a monocotyledon.

Investigation

Materials

- Two flowers of tulip, labelled A
- Two flowers of polyanthus, labelled B
- Scalpel
- Seeker
- Hand lens

Method

1 You have two flowers, A and B, both of which are pollinated by the same agent.

(a) **Name the agent responsible for pollination.** (1)

(b) **Give two reasons for your answer, referring to features that can be seen in both flowers.** (2)

2 Copy Table 11. **Use the seeker to examine each flower, then complete the table.** (10)

Table 11

Feature	Flower A	Flower B
(a) Petals/perianth segments, joined or free		
(b) Number of stamens		
(c) Region of flower to which stamens are attached		
(d) Ovary, superior or inferior		
(e) Monocotyledon or dicotyledon		

3 Remove three perianth segments and three stamens from flower A. **Make a large, labelled drawing of the half-flower. Label any four of the following structures on your drawing: perianth segment, stigma, style, ovary, receptacle, anther lobes, filament, 'honey guides'.** (5)

4 Cut the ovary of flower A transversely. **Make a drawing to show the number of compartments it contains, together with the attachment and arrangement of the developing seeds.** (2)

Total 20 marks

Taking it further

1 Examine a flower of daffodil. Does it come from a monocotyledon or dicotyledon? How does it differ from tulip? Tabulate (i) similarities and (ii) differences between the two flowers.

24 Wind as an agent of pollination and seed dispersal

TIME
Investigation: 30–40 minutes
July–September

Wind is an agent of cross-pollination, carrying pollen from wind-pollinated flowers to others of the same species. It is also an effective agent of seed dispersal. Wind blows seeds from the plants on which they grew, and carries them to distant places where they have a better chance of germination and survival.

The aim of this investigation is to examine wind-pollinated flowers, and fruits containing wind-dispersed seeds, observing and recording their adaptations to dispersal by wind.

Investigation

Materials

- Inflorescence of grass, labelled A
- Inflorescence of plantain, labelled B
- Fruit of sycamore or maple, labelled C
- Fruit of poppy, labelled D
- Hand lens
- Mounted needle

Method

1 Use the the mounted needle and hand lens to examine one of the grass flowers. **Name any three parts labelled A–H in Figure 26.** (*3*)

2
(a) Both grass (A) and and plantain (B) are pollinated by wind. **Examine the flowers carefully and list three features, shown by both flowers, that are typical of wind-pollinated species.** (*3*)

(b) **What part does the flower stalk of grass (A) and plantain (B) play in helping to disperse pollen?** (*2*)

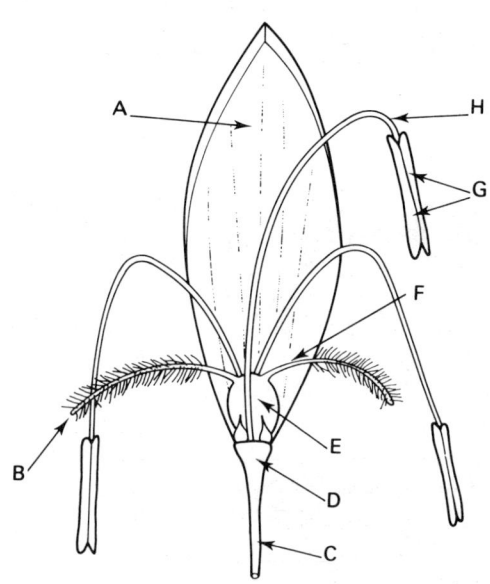

Fig. 26 *Flower of grass*

(c) **In the plantain (B) do the stamens or carpels of the flower ripen first?** (*1*)

(d) **What is the advantage of male and female parts of a flower ripening at different times?** (*1*)

3
(a) **Make a large, labelled drawing (×1) of the sycamore fruit (C), to show how it hangs from the tree. Label any four of the following features: flower stalk, ovary wall, remains of style and stigma, position of seed, petal, and sepal scar.** (*4*)

(b) Stand on the floor, on a chair if your teacher gives permission. Throw the complete sycamore fruit (C) about 1 metre into the air and watch it fall to the ground. Separate the fruit into two halves, and throw one-half about 1 metre into the air, watching it as it falls. **What did you observe as the half-fruit fell?** (*1*)

(c) **What is the advantage to the plant of this movement in the half-fruit?** (*2*)

4

(a) Examine the fruit of poppy (D), known as a capsule. **Describe how seeds are shed from the fruit.** (2)

(b) Inside the capsule there are vertical partitions, like the walls of small rooms, separating seeds in one compartment from those in another. **What is the effect of these partitions on the direction of seed dispersal?**

Total 20 marks

Taking it further

1 Make a list of plants from the herb, shrub, and tree layers of a wood, that are (a) wind pollinated, and (b) have their seeds dispersed by wind. In which layer of vegetation do you find most plants of each type?

25 Distribution of seeds around a tree

TIME
Preparation: 30–40 minutes
Investigation: 30–40 minutes
September–October

In the autumn, wind scatters the seeds of some woodland trees, such as sycamore, lime, and elm. The direction in which the majority of seeds are dispersed, and the distances to which they are carried, can be found by counting seeds that have fallen to the ground. Lengths of coarse string, called transect lines, are pegged out on the ground, spread out from the base of a tree trunk. A counting frame, or quadrat, is put at evenly spaced points along each line. Counts are made of the number of seeds on the ground at each point. After comparing results from different positions around the tree, it is possible to tell if seeds are dispersed evenly around the tree, or mostly in one direction.

Investigation

Materials

- Four transect lines, marked at distances of 5 m
- m² quadrat
- Eight pages
- Compass

Method

1 Peg out four transect lines to spread outwards from the base of a large tree. Arrange the transect lines at right angles to one another, so that they respectively face north, south, east, and west. See Figure 27.

Fig. 27 *Lay out of the transect lines*

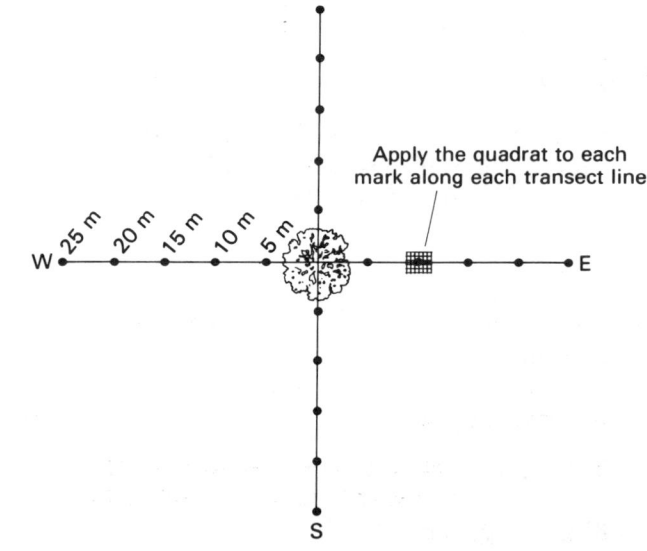

48

2 Copy Table 12. Put the quadrat on one of the transect lines, touching the base of the tree Count and record the number of seeds in the quadrat. Repeat counts every 5 m along the transect line, until you have reached a distance of 25 m from the tree. Record your results. Make further counts, at distance of 5 m, along the other transect lines. Record your results.

Table 12

Direction	Distance from tree trunk (m)					
	0	5	10	15	20	25
North (N)						
South (S)						
East (E)						
West (W)						

Investigation

Materials

- Table of results
- Graph paper

Method

1 **Make a neat copy of the table showing your results.** *(4)*

2 **What was the total number of seeds counted along lines facing in each of the following directions?**

(a) **North** *(1)*

(b) **South** *(1)*

(c) **East** *(1)*

(d) **West** *(1)*

3 **In which direction were:**

(a) **most seeds dispersed;** *(1)*

(b) **least seeds dispersed?** *(1)*

4 **Suggest a reason for any unequal distribution of seeds around the tree.** *(1)*

5 **Draw a graph to show how the number of seeds, on the east-facing side of the tree, varied with distance from the trunk.** *(5)*

6 **Suggest two reasons for the pattern of seed dispersal** *(2)*

7 **Name two further aspects of seed dispersal from a tree that could be investigated.** *(2)*

Total 20 marks

Taking it further

1 Seeds of some trees such as oak and horse chestnut have no special features that assist their dispersal by wind. Does wind affect the direction in which these seeds are distributed, or are they distributed evenly around the tree from which they fell?

2 The longer a seed takes to fall from a tree, the better are its chances of being blown by the wind. Fruits of lime are said to fall more rapidly than fruits of sycamore. Design an experiment to test this hypothesis.
 (a) State your hypothesis.
 (b) How would you test your hypothesis?
 (c) What do you predict would happen if your hypothesis was correct?
 (d) Draw a table, with headings, for your results.

Ripe fruits of sycamore, ready for dispersal by wind.

26 Depth of sowing and its effects on germination

TIME
Preparation: 15–20 minutes
Observation: 3–6 weeks
Investigation: 30–40 minutes
April–September

Seeds of French bean require moisture, oxygen, and a suitable temperature before they will germinate. When these seeds are sown in the soil, other factors may affect the number of seeds that germinate, and the rate of growth of the seedlings. People who sell seeds often need to know the percentage of germination, and the time taken for germination. Percentage of germination is the number of seeds out of every 100 sown that grow into seedlings. The time taken for germination is more difficult to define, as seedlings grow at different rates, and some seeds may fail to germinate. A common practice is to record the time taken for one-half (50%) of the seedlings to break through the soil surface.

This is an investigation of the effects of different depths of sowing on percentage of germination, and the time taken for shoots to appear above the soil.

Preparation

Materials

- Forty seeds of French bean
- Four flower pots, 15 cm in depth
- John Innes compost
- Ruler, graduated in millimetres
- Glass-marking pen

Method

1 Number the flower pots from 1–4. Put ten beans, evenly spaced, in the bottom of pot 1. Cover the seeds with compost and fill the pot to the brim. (These seeds have been planted at a depth of 15 cm.) See Figure 28.
2 Use the ruler to measure a distance of 5 cm from the bottom of the pot. Mark this point. Put compost into the pot as far as this mark, then sow ten seeds. Fill the pot to the brim with compost.
3 Similarly, sow 10 seeds in pot 3 at 10 cm from the bottom of the pot, and 10 seeds in pot 4 at the surface. Water the pots and put them in a warm, sunny place, either on a shelf in the greenhouse, or inside a polythene bag tied with string.

Fig. 28 *Setting up Investigation 26*
(i) Pot 1

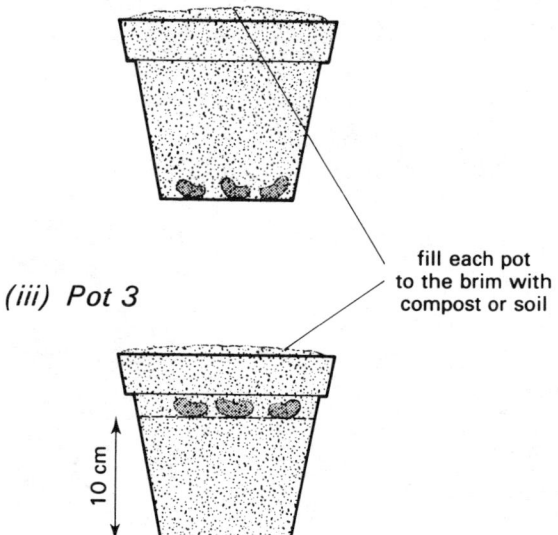

(ii) Pot 2

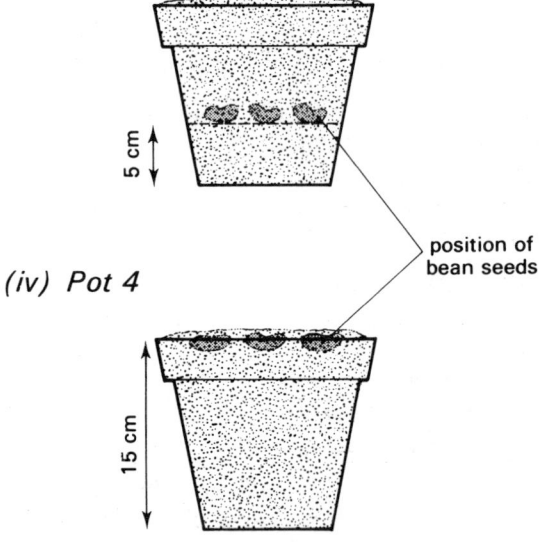

Observation

1. Examine the pots every 2–3 days, and keep them watered.
2. Keep records of both (i) the percentage of germination and (ii) the time taken for 50% of the seedlings to appear above the soil.

Investigation

Materials

- Record of results
- Graph paper

Method

1. **Tabulate your results for:**

   (a) **percentage of germination;** (2)

   (b) **time taken for 50% of the seedlings to appear above the soil.** (2)

2. **Draw graphs to show:** (3)

   (a) **percentage of germination**

   (b) **time taken for 50% of seedlings to appear above the soil.** (3)

3.
   (a) What was the depth of sowing in pot 3? (1)

   (b) What was the optimum depth of sowing? (1)

   (c) What were the advantages of sowing seeds at an optimum depth? (2)

4. **Suggest two reasons why seeds sown at the surface may show a lower percentage of germination than those sown more deeply.** (2)

5. **Suggest two reasons why seeds sown deeply in the soil may show a lower percentage of germination than those sown just below the surface.** (2)

6. Depth of sowing is not the only factor that may prevent seedlings from germinating. **Name two different pests that may reduce the percentage of germination by attacking seeds or seedlings.** (2)

Total 20 marks

Taking it further

1. Some gardeners say it is possible to grow beans from half a seed. Design an experiment to test this hypothesis.

   (a) Describe how you would prepare your half-seeds.

   (b) State your hypothesis.

   (b) How would you test your hypothesis?

   (c) What do you predict would happen if your hypothesis was correct?

   (d) Draw a table, with headings, for your results.

Young seedlings of dwarf french bean

27 Producing new plants from stem cuttings

TIME
Preparation: 10–15 minutes
Observation: 2–4 weeks
Investigation: 30–40 minutes

Many plants reproduce asexually, forming young, independent offspring from their leaves, stems, or roots. Horticulturalists and gardeners make good use of this ability to increase their stocks of desirable flowering plants, fruits, and vegetables. Plants producing beautiful flowers, attractive leaves, sweet fruits, or tasty vegetables can all be increased by asexual reproduction. Commercial growers often produce large numbers of herbs, shrubs, and trees from stem cuttings (short lengths of stem bearing buds and leaves). Successful growth of a stem cutting depends on the formation of roots. Therefore, a rooting hormone (auxin) is usually applied to the base of the cuttings to stimulate root formation.

The aim of this investigation is to discover the effects on root formation in cuttings of (a) auxin, and (b) removing part of the epidermis. The plant used is mint, a garden herb which forms roots when its cut stems are immersed in water.

Preparation

Materials

- Twelve cut shoots of mint, each bearing 10 or more leaves
- Rooting hormone (containing auxin)
- Four 250 cm³ beakers
- Scouring pad, or coarse glass paper
- Glass-marking pen

Method

1 Label the beakers from 1–4. Pour 200 cm³ tap water into each beaker. Mark the water level on the beaker. Add one small drop of rooting liquid, or a knifepoint of rooting powder, to the water in beakers 3 and 4.
2 Select cut shoots of a fairly uniform size. Retain the 10 leaves nearest the apex of each shoot, but remove the other leaves to expose a 5–10 cm length of stem at the base. Place three prepared shoots into beaker 1, and three in beaker 3.
3 Use the scouring pad, or glass paper, to rub away part of the epidermis on the exposed stem of the remaining six cuttings. (A single movement, along each side of the stem, from the top towards the base, should be sufficient.) Place three of these shoots into beaker 2, and three into beaker 4. See Figure 29. Stand all of the beakers on a bench in the greenhouse, preferably in bright light, but shaded from direct sunlight.

Fig.29 *Arrangement of cut shoots of mint*

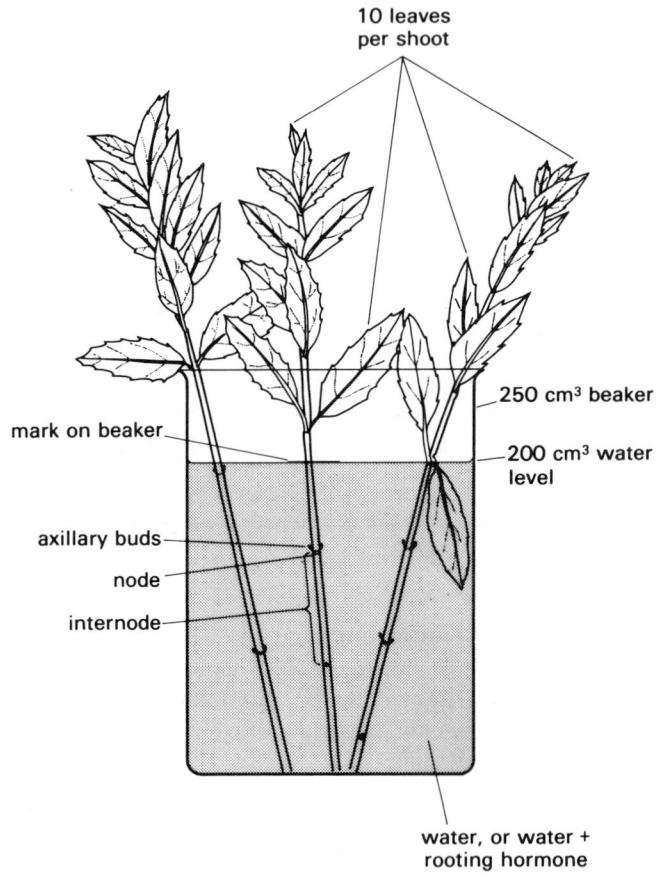

4 Copy Table 13.

Over a period of 30 days make careful daily observations of the cuttings. Record the number of adventitious roots emerging from the three cuttings in each beaker.

Table 13 *Recording growth of roots*

Time (days)	Number of roots formed			
	Beaker 1	2	3	4
0				
1				
2				
3 . . . 28				
29				
30				

Investigation

Materials

Beakers 1, 2, 3 and 4, each with three cuttings of mint
Record of results

Method

1 **Make a neat copy of your table, recording development of adventitious roots in the four different groups of cuttings.** (5)

2 **Which of the beakers contained the following?**
   (a) **water** (1)
   (b) **water and auxin** (1)
   (c) **stems with scratched surfaces** (1)

3 Examine the cuttings that have stood in water.
   (a) **From which part of the stem did adventitious roots appear?** (1)
   (b) **What has happened to axillary buds from which the leaves have been removed?** (1)

4 **Comment on the effects of**
   (a) **adding auxin to the water;** (3)
   (b) **scratching the surface of the stem.** (3)

5 Rhododendrons may be propagated from cuttings in a coarse grit mixed with peat. Take a 15 cm stem cutting and strip off the leaves from the base of the stem, leaving only the four terminal leaves attached. Dip the cut end of the stem into rooting powder. Push the cut end of the stem into the soil, to a depth of about 10 cm. Put the pot inside a polythene bag. Blow expired air into the bag until it is inflated, then tie it with string. See Figure 30. Suggest reasons for each of the following, when preparing the cutting:

Fig. 30 *Propagating a stem cutting of rhododendron*

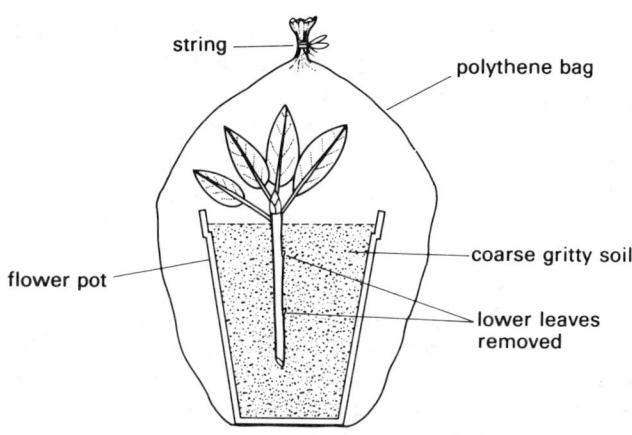

(a) **removing most of the leaves;** (1)
(b) **pushing the stem through coarse grit;** (1)
(c) **placing the pot inside a polythene bag;** (1)
(d) **blowing expired air into the bag.** (1)

Total 20 marks

Taking it further

1 A fungicide called Benlate, which is available from garden centres, can be used to assist the rooting of cuttings. It does so by preventing the base of the stems from rotting. Design an experiment to find out if Benlate helps to promote the rooting of rhododendron cuttings.
   (a) State your hypothesis.
   (b) How would you test your hypothesis?
   (c) What do you predict would happen if your hypothesis was correct?
   (d) Draw a table, with headings, for your results.

28 Comparing rates of growth in parts of a seedling

TIME
Preparation: 5–10 minutes
Observation: 7–10 days
Investigation: 30–40 minutes

Germination is the growth of a seed into a seedling. Seeds contain embryo plants and a supply of food. During germination, food passes into the embryo plant, causing it to increase in size and grow from the seed.

Maize is a convenient plant to study, because it has a large grain, from which a single root (radicle) and a single shoot (plumule) appear. See Figure 31. If seedlings are grown in a dish, supported by a clear jelly, all parts of the seedling are visible. Moreover, adhesive measuring tape can be stuck to the outside of the dish, making it possible to record changes in the lengths of parts of the seedling as they grow. The aim of this investigation is to observe the process of germination in maize, and to measure, record, and compare rates of growth in the plumule and radicle.

Fig. 31 *Young seedling of maize*

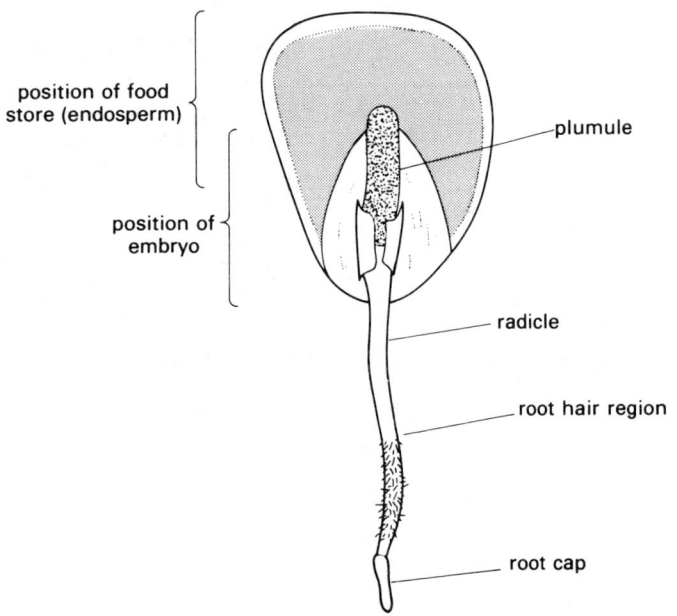

Preparation

Materials
- Petri dish, containing four maize grains in agar
- Small flower pot or adhesive tape

Method
1 You are provided with a petri dish, containing four maize grains in agar. Select a warm place, in light or darkness, where the dish can remain for several days. Support the dish in a vertical position, either in a small flower pot, or held against a wall by two pieces of adhesive tape. See Figure 32.

Fig. 32 *Methods of supporting the petri dish in a vertical position*

(i) The dish is placed inside a small flower pot

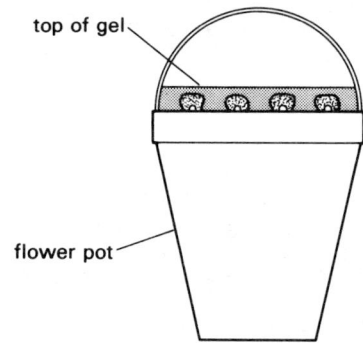

(ii) The dish is suspended, by means of strips of adhesive tape, from a wall

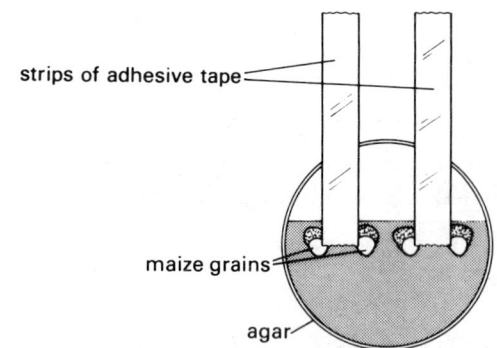

Observation

Materials

- Scalafix tape
- Ruler, graduated in millimetres
- Marking ink, and pen

Method

1. Examine the dish daily, concentrating on one or two seedlings that are growing rapidly. Record the day on which the (i) radicle, and (ii) plumule appeared from the grain.
2. As soon as the young root (radicle) has appeared, stick a piece of Scalafix tape to the lid of the petri dish, parallel to the root. Measure and record the daily increase in the length of the root over the next three days.
3. Similarly, as soon as the young shoot (plumule) appears, measure and record its daily increase in length over the next three days.
4. Select a maize grain with a young root of about 1.5 cm. Make horizontal ink marks 2 mm apart along the entire length of the root. Return the dish to its support, and examine the dish after two days. Write down your observations.
5. Select another maize grain with a young root of about 1.5 cm. Mark it as before, but turn the petri dish on its side, so that this root is horizontal. Examine the dish after two days, and write down your observations.

Investigation

Materials

- Record of results

Method

1
  (a) **What was the first structure to grow from the maize grain?** *(1)*
  (b) **In which direction did it grow?** *(1)*
  (c) **On which day did it first appear?** *(1)*

2
  (a) **What was the second structure to grow from the maize grain?** *(1)*
  (b) **In which direction did it grow?** *(1)*
  (c) **On which day did it first appear?** *(1)*
  (d) **What external force determines the direction in which the young root and shoot grow?** *(1)*

3
  (a) **Tabulate your results for root growth.** *(2)*
  (b) **Describe the pattern of growth of the root.** *(2)*

4
  (a) **Tabulate your results for shoot growth.** *(2)*
  (b) **How does the rate of growth of the young shoot compare with that of the young root?** *(1)*
  (c) **Suggest a reason for any differences you have observed?** *(2)*

5. You drew ink marks 2 mm apart on a root growing vertically. **Describe what you saw after 2 days, and explain your observation.** *(2)*

6. You marked another root, then placed in horizontally. **Describe what you saw after 2 days, and explain your observation.** *(2)*

Total 20 marks

Taking it further

1. Design an experiment to find out if seedlings of wheat grow faster than those of barley.
   (a) List the materials and apparatus for your experiment.
   (b) Write out your method, listing instructions in the order they should be carried out.
   (c) Draw a table, with headings, for your results.
2. A scientist claims that the roots of maize seedlings grow faster at night than during the day. Design an experiment to test this hypothesis.
   (a) State your hypothesis.
   (b) How would you test this hypothesis?
   (c) What do you predict would happen if your hypothesis was correct?
   (d) Draw a table, with headings, for your results.

29 Tropic responses and auxin production

TIME
Preparation: 15–20 minutes
Observation: 7–10 days
Investigation: 30–40 minutes

The shoots and roots of seedlings are sensitive both to gravity and to the direction of light. If seedlings are illuminated from one side, shoots bend towards the light source and most roots bend away from it. The response is called phototropism, a growth movement caused by different intensities of light on opposite surfaces of a shoot or root. Shoots make a positive response by growing towards the light source. Roots make a negative response by growing away from it.

Growth, including the tropic response to light, is controlled by a plant hormone called auxin, produced by the tip of the shoot. In this investigation you find out the effects of the position of a light source on the growth of seedlings, and consider evidence for the presence of a growth-stimulating substance in the tip of the shoot.

Preparation

Materials

- Thirty barley grains
- Petri dish, containing four or more germinating maize grains
- Shoe box
- Three petri dish bases
- Small flower pot or adhesive tape
- Tissue paper
- Electric lamp, fitted with a 60 W bulb
- Scissors
- Ruler, graduated in millimetres
- Hand lens
- Scalpel
- Pencil
- Glass-marking pen

Method

> **SAFETY PRECAUTIONS**
>
> *You have a sharp scalpel. Handle it with care.*

1 Remove the lid of the shoe box. Draw lines across it from corner to corner to find the central point. Use the scissors to cut a hole 7 × 7 cm at the centre.

2 Number the petri dish bases from 1–3. Put two paper tissues into the base of each petri dish and add sufficient water to moisten the paper. Drain off any excess water, and sow 10 barley grains per dish. Put the three dishes into the base of the shoe box, and arrange them as shown in Figure 33.

Fig. 33 *Arrangement of dishes in the shoe box*

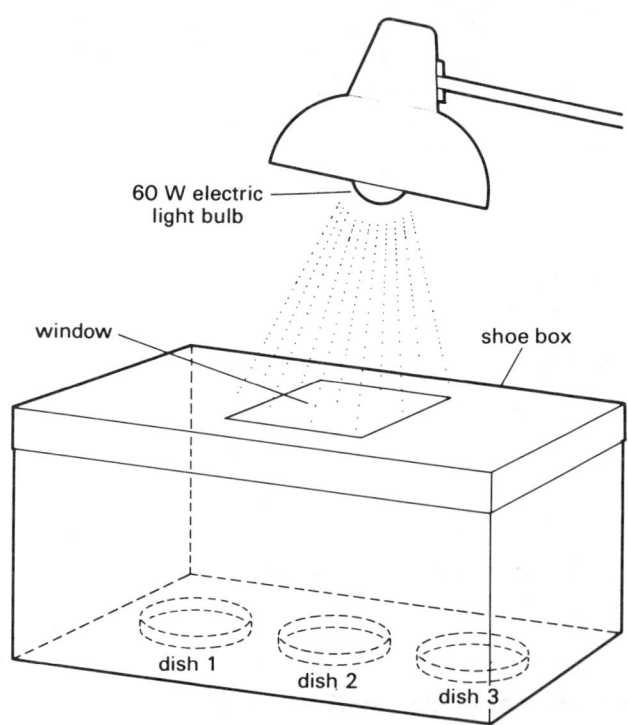

Place the box in a brightly lit greenhouse, or beneath an illuminated 60 W electric light bulb. Keep the dishes moist by adding water whenever they begin to dry out. Allow 7–10 days for the barley grains to germinate.

3 You are provided with a petri dish containing four or more germinating maize grains. Take a hand lens in one hand, and a scalpel in the other, and very carefully remove small pieces of tissue from the tops of three plumules, as shown in Figure 34. Leave at least one plumle tip intact. Replace the lid and support the dish in a vertical position, using either a small flower pot or two strips of adhesive tape (see Figure 32). Place the dish in a warm place and allow 5–10 days for further growth of the plumules.

Fig. 34 *Treatment of maize shoots*

(i) Tip intact   (ii) Tip removed

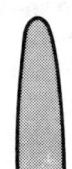 

(iii) Left-half of tip removed   (iv) Right-half of tip removed

Investigation

Materials

- Shoe box, containing germinating barley grains
- Petri dish, containing germinating maize grains

Method

Part A

1 Look at the petri dish bases in the shoe box. **Which dish, or dishes, have received:**

(a) **unilateral (unidirectional) lighting;** *(1)*

(b) **uniform lighting?** *(1)*

2 **Which dish has received the highest light intensity?** *(1)*

3

(a) **How does the colour of seedlings in dish 2 differ from those in dishes 1 and 3?** *(1)*

(b) **Suggest one or more reasons that might explain these differences** *(2)*

4

(a) **How do seedlings in dishes 1 and 3 differ in shape from those in dish 2?** *(1)*

(b) **Suggest a reason for this difference in shape.** *(2)*

(c) **What is the advantage to the seedling of this response?** *(1)*

5

(a) **Do shoots grow more rapidly in light or in darkness?** *(1)*

(b) **On which side do shoots grow most rapidly, the 'dark' side or the 'light' side?** *(1)*

(c) **What happens to the shoots when lighting is uniform?** *(1)*

Part B

6 Look at the maize shoots in the petri dish.

(a) **Which shoot shows most growth?** *(1)*

(b) **What is the general effect of removing the tip?** *(1)*

(c) **What does this suggest about the function of the tip?** *(1)*

(d) **What is the effect of removing one-half of the tip?** *(2)*

(e) **Suggest a reason for the response of shoots from which half of the tip has been removed.** *(2)*

Total 20 marks

Taking it further

1 It is claimed that the shoots of maize bend more in response to unidirectional (unilateral) blue light than to red light. How would you test this hypothesis, using blue and red sheets of plastic as light filters?

(a) State your hypothesis.
(b) How would you test your hypothesis?
(c) What do you predict would happen if your hypothesis was correct?
(d) Draw a table, with headings, for your results.

30 Differences between seeds and seedlings

TIME
Investigation: 30–40 minutes

There are two major groups of flowering plants, the dicotyledons and monocotyledons. French bean is a dicotyledon. Bean seeds contain two thick, fleshy seed-leaves, or cotyledons, in which food is stored. Maize is a monocotyledon. There is a single, small cotyledon in this seed, and a special storage tissue for food, the endosperm.

When the seeds germinate, the two cotyledons of French bean are raised above ground. This is epigeal germination. Both cotyledons may turn green and carry out photosynthesis, until the first foliage leaves are able to carry out that role. The maize grain remains below ground during germination. A single shoot emerges above the soil surface, and a protective cover, the coleoptile, breaks as the leaves increase in size and begin to unfold. This is hypogeal germination.

In this investigation you are asked to observe and record differences in structure between seeds and seedlings of French bean and maize.

Investigation

Materials

- Two seeds of French bean, labelled A
- Two maize grains, labelled B
- Seedling of French bean, labelled C
- Seedling of maize, labelled D
- Hand lens
- Forceps
- Scalpel

Method

1 You have French bean seeds (A) and maize grains (B). **Make large, labelled drawings to show the external appearance of both specimens. Label any three of the following structures: hilum, micropyle, embryo, of radicle, remains of style and stigma.** (3)

2 Split the testa of one of the French bean seeds (A). Carefully separate the two cotyledons, so that they lie on the bench surface, surrounded by the testa. Cut one of the maize grains vertically into two equal pieces, passing through the middle of the embryo. **Make large, labelled drawings (× 3) to show the internal structure of the French bean seed and maize grain. Label any three of the following structures on your drawings: endosperm cotyledon, cotyledon stalk, plumule, radicle.** (5)

3 **Make large, labelled drawings (× 3) of the seedlings of French bean (C) amd maize (D). Label any three of the following structures on your drawings: primary root, secondary roots, foilage leaves, cotyledons, coleoptile.** (6)

4 **State two differences between the seedling of French bean (C) and maize (D).** (4)

5 **What is the chief function of the cotyledons of French bean in each of the following stages?**
(a) Seed stage (1)
(b) Seedling stage (1)

Total 20 marks

Taking it further

1 Some gardeners sow seeds in boxes, then transplant young plants to open ground in the garden. It is claimed that if seeds of plants such as maize are sown directly into open ground, they eventually produce larger plants than box-grown seedlings. How could you find out if this claim is correct?

(a) State your hypothesis.
(b) How would you test your hypothesis?
(c) What do you predict would happen if your hypothesis was correct?
(d) Draw a table, with headings, for your results.
(e) Suggest three factors that might cause seeds sown in open ground to produce larger plants.

31 Adaptation: form and function in plant stems

TIME
Investigation: 30–40 minutes
November–April

A plant stem bears buds and leaves. The terminal bud, at the end of the stem, is responsible for growth in length. The axillary buds, in the axils of leaves, develop into branches. Nodes are points on the stem from which the leaves arise. The spaces between the nodes are called internodes. See Figure 35.

According to the theory of evolution, all living things have common ancestry. Common features are shared by organisms that are closely related. This means that those parts of plants that have developed from stems usually bear buds and leaves. At the same time, structures of similar origin may become changed in form (adapted) to carry out different functions. In addition to supporting the leaves, and conducting foodstuffs and water, the stem may serve as a storage organ, carry out asexual reproduction, and make food.

You have three specimens labelled A, B and C. Look at them carefully and try to explain their similarities and differences in terms of the theory of evolution.

Investigation

Materials

- Woody deciduous twig, labelled A
- Brussels sprout, labelled B
- Potato, labelled C
- Hand lens
- Scalpel

Method

1 **Make a large drawing of specimen A, not less than 10 cm in length. Label the following features on your drawing: terminal bud, axillary bud, node, internode, leaf scar.** (4)

Fig. 35 *Generalised structure of a stem*

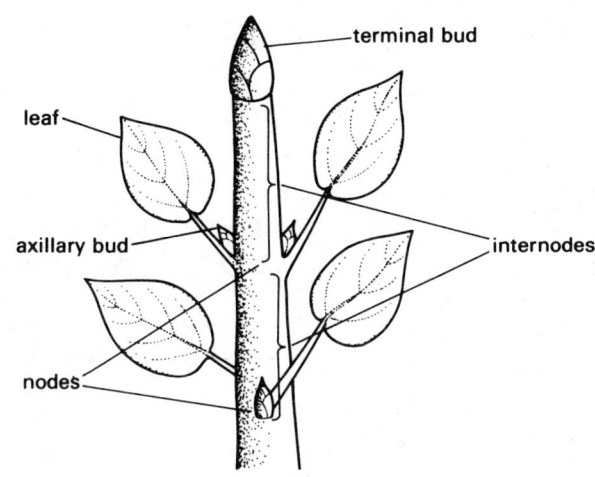

2 Put specimen B on its side. Cut downwards with the scalpel to bisect the specimen. **Make a drawing of the cut surface, not less than 5 cm in length. Label any three features that occur in stems.** (4)

3 **List two similarities between specimens A and B.** (2)

4 **List two differences between specimens A and B.** (4)

5
(a) **From which part of the parent plant has specimen C been formed?** (1)

(b) **What external features of specimen C help to support the answer you have given?** (2)

6 **Comment on how specimens A, B and C help to illustrate the principle of evolution.** (3)

Total 20 marks

Taking it further

1 Collect flowers from five different plants belonging to the same family. List the similarities and differences between them. How do biologists explain (i) their similarities and (ii) their differences?

32 Behaviour of the woodlouse

TIME
Investigation: 30–40 minutes

A number of invertebrate animals, including woodlice, make automatic responses to external stimuli. They may, for example, move towards or away from stimuli such as light, water, touch, or heat. These automatic responses are called taxes. Movement towards a stimulus is a positive taxis, while that in the opposite direction is a negative taxis. The response to light is called phototaxis. Similarly, hydrotaxis is the response to water, and thrigmotaxis the response to touch, or contact with solid objects.

You are asked to investigate some aspects of behaviour in the woodlouse, an animal found living in damp places, beneath stones or rotting wood.

Investigation

Materials

- Four living woodlice in a stoppered boiling tube
- Plastic petri dish
- Protractor
- Black paper/plastic sheet
- Drying agent, in a piece of nylon
- Bench lamp, fitted with a 60 W bulb
- Moist cotton wool
- Elastic band
- Forceps and hand lens
- Glass-marking pen

Method

1 Examine the woodlice in the tube. Copy the outline dorsal view of a woodlouse shown in Figure 36.

(a) **Add to this outline two named organs that respond to external stimuli.** (2)

(b) **To which stimuli do each of these organs respond?** (2)

2 Put the sheet of black paper/plastic over one-half of the boiling tube to form a sleeve. Hold it in place with the elastic band. Rest the tube horizontally on the bench surface, as shown in Figure 37. Switch on the bench lamp and put it at 10 cm distance from the tube.

(a) **How did the woodlice respond to the light?** (1)

(b) **What name is given to this behaviour?** (1)

(c) **Describe two ways in which woodlice might benefit by making this response.** (2)

Fig. 36 *Dorsal view of a woodlouse (Oniscus sp.)*

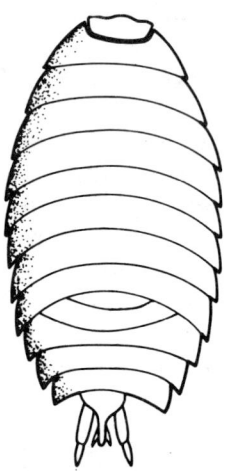

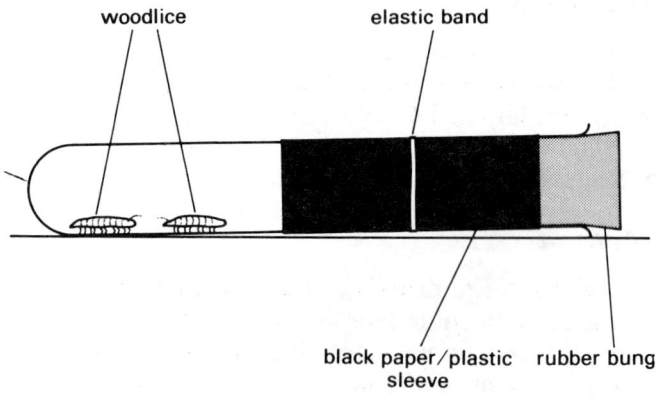

Fig. 37 *Arrangement of the tube to investigate response to light*

3 Remove the elastic band and black paper/plastic sleeve from around the boiling tube. Use forceps to put the piece of nylon containing a drying agent (calcium chloride or silica gel) at the bottom of the boiling tube. Rest the tube horizontally on the bench surface. Put a piece of moist cotton wool at the opposite end, beneath the bung. Use the marking pen to draw a vertical line, dividing the tube into 'dry' and 'wet' halves, as shown in Figure 38.

Fig. 38 *Arrangement of the tube to investigate response to water*

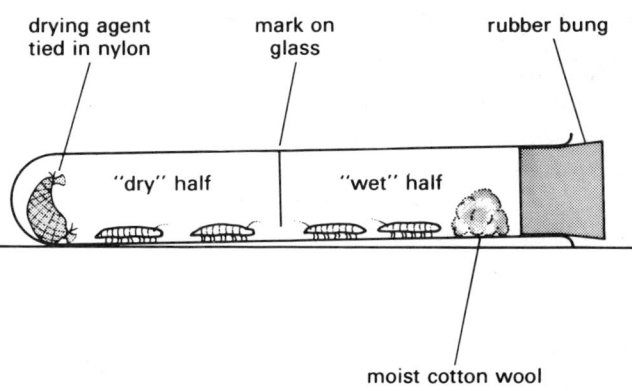

(a) **Explain why the drying agent was put at the bottom of the tube, and not next to the bung.** *(1)*

(b) **Do the woodlice move into the 'wet' or 'dry' half of the tube?** *(1)*

(c) **Give the correct scientific name for the response shown by the woodlice.** *(1)*

4 Remove the woodlice from the boling tube and put them into the petri dish. Bring the bench lamp close to the petri dish.

(a) **How do the woodlice respond?** *(2)*

(b) **To what are the woodlice responding?** *(2)*

(c) **How might the woodlouse benefit by behaving in this way?** *(2)*

5 Suppose you were given a piece of plasticine (or Blu-tack) and a protractor. Use these items, and a single woodlouse in a boiling tube, to design as experiment to find out if a woodlouse can climb up the glass side of a boiling tube, inclined at an angle 20° to the bench surface. **Make a labelled drawing to show the arrangement of the tube and plasticine, (or Blu-tack). Show the position of the woodlouse at the beginning of your experiment.** *(3)*

Total 20 marks

A common woodlouse (Oniscus ascellus)

Taking it further

1 Design an experiment to find out if woodlice move faster in a moist atmosphere than in a dry one. A woodlouse can be made to walk in a straight line by putting it between two parallel glass rods, spaced about 1.0 cm apart.
  (a) List the materials and apparatus for your experiment.
  (b) Write out your method, listing instructions in the order they should be carried out.
  (c) Draw a table, with headings, for your results.
2 It is said that woodlice can climb up a steeper slope than blowfly maggots. Design an experiment to test this hypothesis.
  (a) State your hypothesis.
  (b) How would you test your hypothesis?
  (c) What do you predict would happen if your hypothesis was correct?
  (d) Draw a table, with headings, for your results.

33 Body size in mammals

TIME
Investigation: 50–60 minutes

Suppose the body of a mammal weighs 100 g and is represented by a cube, with sides of 5 cm. Then:
  volume of body = $5 \times 5 \times 5 = 125\,\text{cm}^3$
  surface area of body = $5 \times 5 \times 6 = 150\,\text{cm}^2$
  mass of body = 100 g.

If the animal grows to twice its original size, then:
  volume of body = $10 \times 10 \times 10 = 1{,}000\,\text{cm}^3$
  surface area of body = $10 \times 10 \times 6 = 600\,\text{cm}^2$
  mass of body = $100 \times 8 = 800\,\text{g}$

If the length and height of the body is doubled, the surface area is increased by a factor of four ($\times 4$). Mass and volume are increased by a factor of eight ($\times 8$). The relationship between surface area and mass sets a limit to the size of mammals. See Figure 39. As mammals are warm-blooded, heat is lost via the surface of the body. The larger the ratio, body-surface : volume (or mass), the more rapidly will heat be lost. Again, as body-mass increase, limb bones become more heavily loaded.

The aim of this investigation is to discover the effects of size, mass and surface area on heat loss, and to test the strength of supporting materials.

Investigation

Materials

- Transverse section limb bone
- 500 cm³ beaker
- 250 cm³ beaker
- 100 cm³ beaker
- Two thermometers
- Six sheets of M or foolscap paper
- Packets of exercise books (or alternative means of loading)
- Bunsen burner, tripod, and gauze
- Ruler, graduated in millimetres
- Scissors
- Adhesive tape

Fig. 39 *Scaling up the size of an animal by a factor of two*

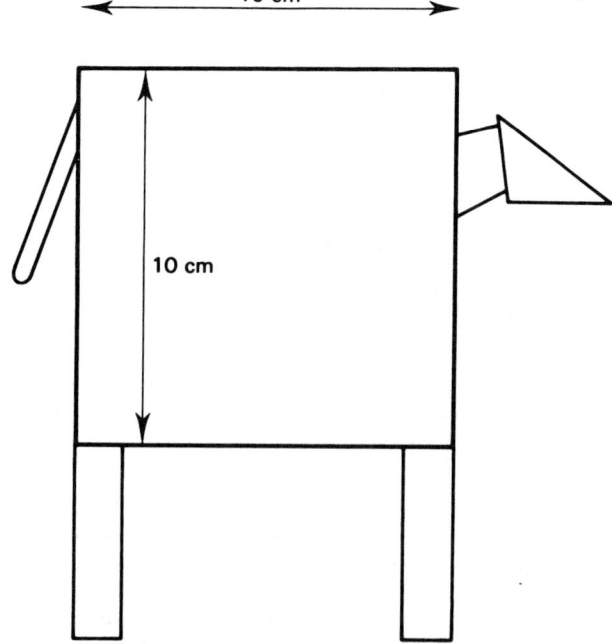

Surface area : volume ratio
 = $\frac{600}{1{,}000}$ = 0.6

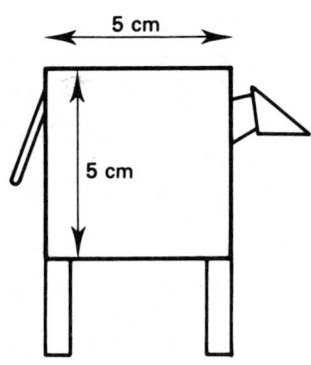

Surface area : volume ratio
 = $\frac{150}{125}$ = 1.2

Method

Part A

1 Fill the 500 cm³ beaker with water to the 400 cm³ mark, and heat the water to 70° C. Carefully pour this water into the smaller beakers, to fill one to the 100 cm³ mark and the other to the 250 cm³ mark. Put a thermometer into each beaker.

(a) **At intervals of 5 minutes, over a period of 30 minutes, read each thermometer and record the temperatures. Tabulate your results.** (3)

(b) **Draw a graph of your results.** (4)

(c) **Which beaker loses heat most rapidly?** (1)

(d) **What is the reason for the greater heat loss from one of the beakers?** (1)

(e) **Would a naked baby or a naked adult lose heat most rapidly? Give a reason for your answer.** (2)

Part B

2 Roll up four sheets of paper to form cylinders with a diameter of approximately 2–3 cm. Use adhesive tape to prevent the cylinders from unrolling. Stand the cylinders on their ends, and arrange them to form the corners of a small square. Carefully load them with books of a similar size, such as exercise books, until the pile collapses.

(a) **How many books formed the load when the columns collapsed?** (If you have an alternative system for loading, you will be shown how it is used.) (1)

(b) Prepare two more similar rolls, then cut each in half. **How many books can be supported by these shorter cylinders before the columns collapsed?** (1)

(c) **In mammals, which structures carry out the same function as the paper cylinders?** (1)

(d) **To which force were the paper cylinders subjected? (i) extension, (ii) compression, (iii) tension, (iv) flexion?** (1)

(e) **Are short, broad cylinders, or long, narrow ones best able to resist this force?** (1)

3 Look at the section of a limb bone.

(a) **In what way does it resemble one of the paper cylinders?** (1)

(b) **In what way does it differ from a paper cylinder?** (1)

4 Overweight (obese) people often experience discomfort when running for more than 50 metres, and they rarely win races. **Suggest reasons for this.** (2)

Total 20 marks

Taking it further

1 Height and body-mass in humans have a marked effect on athletic performance. For example, the best high-jumpers are usually tall and of relatively low mass. Carry out the following investigation to find out if height and body-mass have any effect on performance in the high-jump.

(a) Form a group of 10 pupils.
(b) Measure and record the height (cm) and mass (kg) of each member of the group.
(c) Use the following formula to express the relationship between the height and mass of each person:

$$\text{height/mass index} = \frac{\text{height (cm)}}{\text{mass (kg)}}$$

(d) Draw a bar graph of your results.
(e) Measure the maximum height to which each member of the group can jump.
(f) Draw a bar graph of your results.
(g) What is the general relationship between the height/mass index, and height jumped?

Gerd Wessig, an international high-jumper. How could you measure and record an athelete's performance in the high jump?

34 Exercise: measuring performance

TIME
Preparation: 30–40 minutes
Investigation: 30–40 minutes

Many people take an active part in sports and games. These pleasant, sociable activities can improve our health and help us to remain active into old age. Taking regular exercise has a number of beneficial effects:

1. All movement requires energy. During exercise, carbohydrates that might otherwise be converted to fat are used up.
2. Running, swimming and other forms of vigorous exercise strengthen muscles, so that they tire less easily.
3. Any exercise that increases the rate of heartbeat improves the efficiency of the heart and lungs. The heart responds to regular vigorous exercise by pumping more blood to the tissues.
4. Using muscles helps to keep the joints in a supple condition.

Performances in almost any exercise can be recorded scientifically. Such records may be used by athletes to improve their own performances, or for comparison with other athletes. This investigation shows some of the ways in which the performance of your body can be recorded. You are asked to record the rate at which you run 20 M stretches of a 100 M race, measure changes in your pulse rate during physical exercise and devise a method for measuring flexibility of the joints in the legs and back.

Preparation

Materials

- Gymnasium bench
- Metronome
- Five stop-watches
- Wrist watch

Method

> **SAFETY PRECAUTIONS**
>
> *Do not strain while carrying out these exercises, as this may cause injury.*
> *Do not attempt any of the exercises against professional medical advice.*

1. You are asked to run a timed race over 100 metres. Find six friends who are willing to help. Ask one to act as the starter by waving a white handkerchief. Give stop-watches to the others and ask them to take up positions at distances of 20 m along the track. As you pass each recorder, a stop-watch is pressed, to record your time at that distance. Copy Table 14. Tabulate your results.

Table 14

Distance (m)	Time (s)	Time for each 20 m stretch (s)
20		
40		
60		
80		
100		

2. Ask permission to use a bench from the gymnasium. Set the metronome at 30 beats per minute. Measure and record the pulse rate at your wrist before you begin to exercise. Step on-and-off the bench 30 times a minute, in time with the metronome. In each movement lead with one foot, place both feet on the bench, return one foot to the floor, then both feet. Continue with the exercise for 3 minutes. Record your pulse rate immediately after exercising, and at intervals of one minute until your pulse returns to its normal rate. Make a record of your results.

Investigation

Materials
- Record of results
- Graph paper

Method

Part A

1

(a) **Make a neat copy of the table showing your times during the 100 m run.** (2)

(b) **Plot your results as a bar graph.** (2)

(c) **Over which part of the course were you running fastest?** (1)

(d) **Calculate your speed over this distance in metres per hour. Show how you arrive at your answer.** (2)

Part B

2

(a) **Draw a graph of your rate of heartbeat (pulse rate) before, during and after the stepping exercise.** (4)

(b) **What does the rate of heartbeat during exercise tell us about the condition of the heart?** (1)

(c) **If you repeated the exercise several months after your first attempt, how would you explain the following results?**

 (i) **The maximum rate of heartbeat was slower.** (1)

 (ii) **The maximum rate of heartbeat was faster.** (1)

Part C

3 Some people can put the palms of their hands on the ground in front of their feet, while others can barely reach below their knees. Suppose you were asked to produce a scale of flexibility from 1 (= wrist level with knee) to 10 (= wrist in contact with the floor). See Figure 40. **What scale would you use, and how would you measure flexibility of the joints in another person?** (6)

Total 20 marks

Fig. 40 *Flexibility of the legs, back and arms*

(i) Poor flexibility

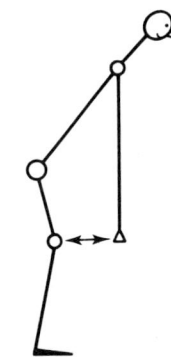

Score = 1 point
the wrist is level with the knee

(ii) Good flexibility

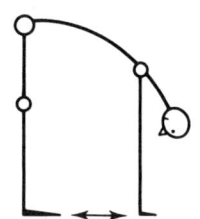

Score = 10 points
the wrist is in contact with the ground

Taking it further

1 Muscles become stronger and more efficient as a result of regular exercise. For example, the more you practice press-ups, the more you should be able to do. Keep a record of your progress. Draw a graph of the number of press-ups you can comfortably complete (remember, no straining), against time (days or weeks).

2 Measure and record your pulse rate at hourly intervals during the day. Draw a graph of your results. What do you conclude? How do the following activities affect your rate of heartbeat: (i) walking, (ii) running, (iii) eating, (iv) resting?

3 It has been claimed that people can run faster at 7:00 p.m. than at 7:00 a.m. Design an experiment to test this hypothesis.
   (a) State your hypothesis.
   (b) How would you test your hypothesis?
   (c) What do you predict would happen if your hypothesis was correct?
   (d) Draw a table, with headings, for your results.

35 Carbon dioxide in inspired and expired air

TIME
Investigation: 30–40 minutes

Air is a mixture of gases. One of these gases, carbon dioxide, forms 0.03% of the air by volume. During respiration in plants and animals carbon dioxide is released to the atmosphere. Respiration in which oxygen is used is aerobic. Aerobic respiration may be summed up in the following equation:

$$C_6H_{12}O_6 + 6O_2 \longrightarrow 6CO_2 + 6H_2O + \text{energy}$$
(glucose) (oxygen) (carbon dioxide) (water)

Respiration that occurs in the absence of oxygen is anaerobic. Anaerobic respiration in animals such as humans takes place during vigorous exercise, when glucose is converted to lactic acid.

glucose $\longrightarrow$ lactic acid + a little energy

Later, after exercise, while heavy breathing continues, some of the lactic acid is converted into carbon dioxide and water.

lactic acid $\longrightarrow$ carbon dioxide + water

The aim of this investigation is to compare amounts of carbon dioxide in inspired and expired air.

Investigation

Materials

- Apparatus as shown in Figure 41
- Three test tubes in a rack
- 15 cm³ hydrogen carbonate (bicarbonate) indicator solution
- 5 cm³ dilute hydrochloric acid
- 5 cm³ dilute sodium hydroxide solution
- 5 cm³ plastic syringe
- 20 cm³ glass tube
- Stop-clock, or watch with a second hand
- Glass-marking pen
- Safety spectacles

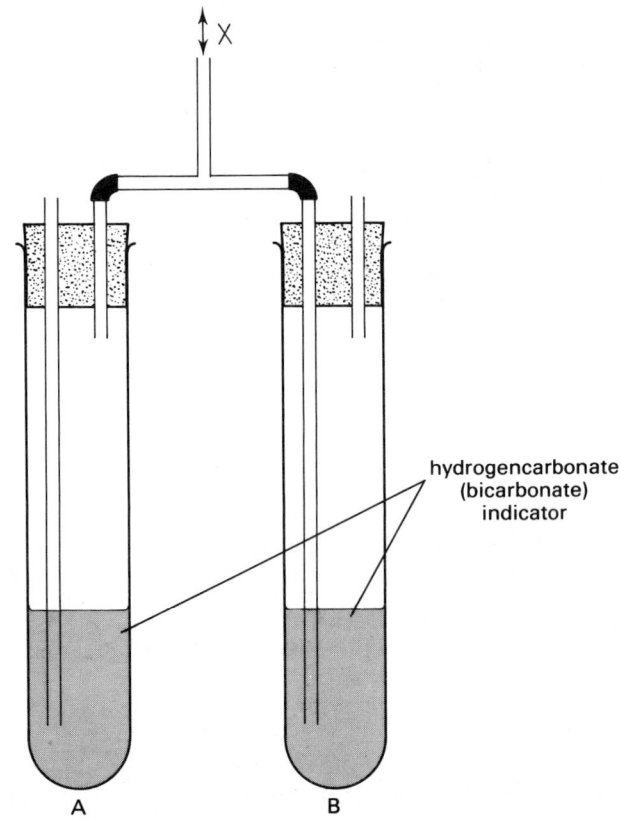

Fig. 41 *Apparatus for comparing amounts of carbon dioxide in inspired and expired air*

Method

SAFETY PRECAUTIONS

Wear safety spectacles. Wash off any spillages with plenty of water.

1. Put on your safety spectacles. Label the test tubes A, B, and C. Use the syringe to put 5 cm³ hydrogen carbonate (bicarbonate) indicator into each tube.

2. Slowly pour the hydrochloric acid into tube A. **What did you observe?** *(1)*

3. Slowly pour the sodium hydroxide solution into tube B. **What did you observe?** *(1)*

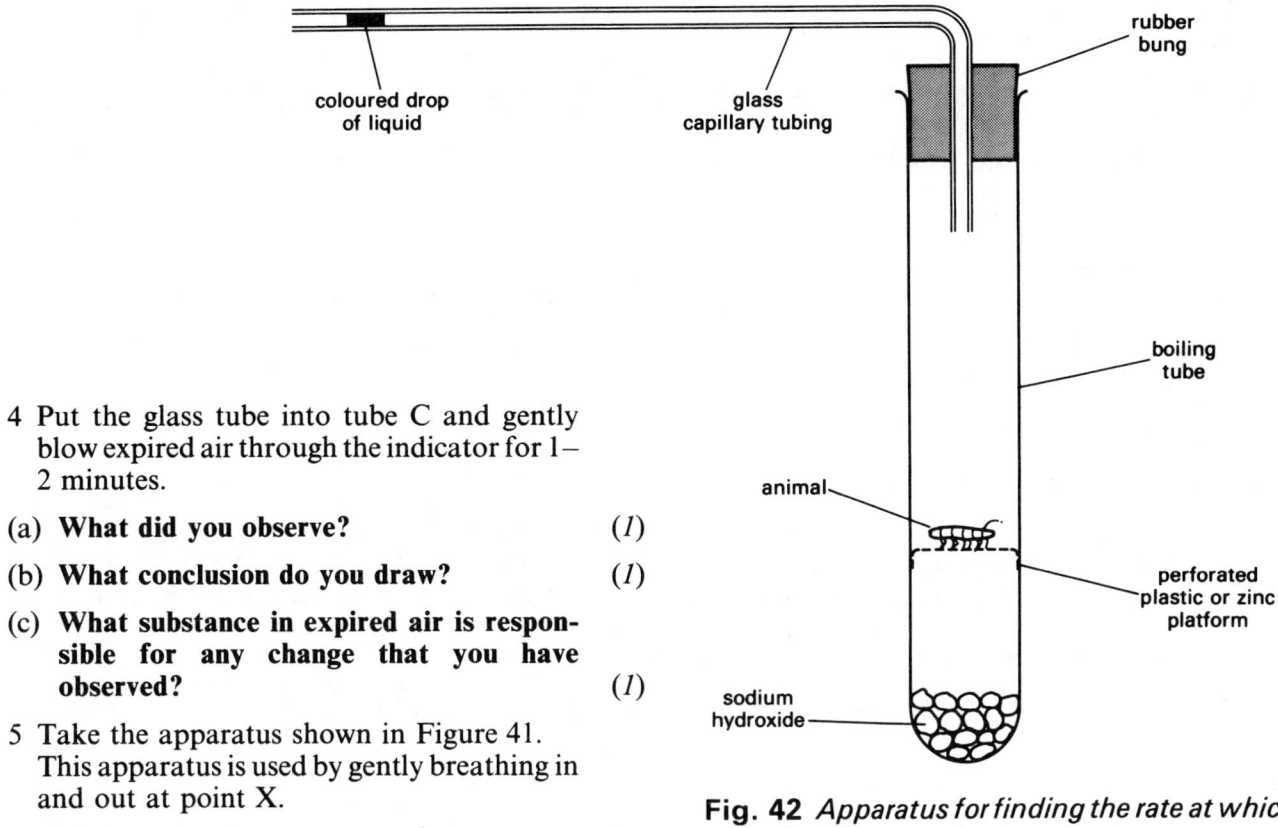

Fig. 42 *Apparatus for finding the rate at which an animal takes up oxygen*

4 Put the glass tube into tube C and gently blow expired air through the indicator for 1–2 minutes.

(a) **What did you observe?** *(1)*

(b) **What conclusion do you draw?** *(1)*

(c) **What substance in expired air is responsible for any change that you have observed?** *(1)*

5 Take the apparatus shown in Figure 41. This apparatus is used by gently breathing in and out at point X.

(a) **Through which tube do you inhale?** *(1)*

(b) **Through which tube do you exhale?** *(1)*

6 Start the stop-clock, or write down the time. Gently breathe in and out through the central tube until the indicator in both tubes changes colour from purple-red to yellow.

(a) **Record the time taken for this colour change in both tubes.** *(2)*

(b) **How do you account for your results?** *(4)*

7 Inspired air contains 0.03% carbon dioxide. **Write an equation to show how you could find the approximate amount of carbon dioxide in expired air.** *(2)*

8 Suppose you left the laboratory, ran around the playing field, and returned to breathe gently in and out of the apparatus.

(a) **How would this affect the rate of colour change in:**

   (i) **tube A** *(1)*

   (ii) **tube B?** *(1)*

(b) **Write a brief explanation of the results that you have predicted.** *(3)*

Total 20 marks

Taking it further

1 The apparatus shown in Figure 42 can be used to find out the rate at which oxygen is taken up by a small animal in a boiling tube. Sodium hydroxide in the bottom of the tube absorbs carbon dioxide. As oxygen is used up, the coloured drop of liquid moves along the capillary, towards the boiling tube.

Design an experiment to find out if a maggot requires more oxygen than a woodlouse.
  (a) List the materials and apparatus for your experiment.
  (b) Write out your method, listing instructions in the order they should be carried out.
  (c) Draw a table, with headings, for your results.

2 Measure and record your rate of breathing at hourly intervals during the day. Draw a graph of your results. What do you conclude? How is your rate of breathing affected by (i) walking, (ii) running, (iii) eating, (iv) resting?

36 Fermentation of glucose by yeast

TIME
Preparation: 10–15 minutes
Observation: 60–80 minutes
Investigation: 30–40 minutes

The process that releases energy in the bodies of plants and animals is called respiration. If oxygen is used, respiration is aerobic. Aerobic respiration may be summarised by the equation:

sugar + oxygen ⟶ carbon dioxide + water + energy

Respiration without oxygen is anaerobic. Anaerobic respiration in yeast results in the formation of ethanol (alcohol).

sugar ⟶ ethanol + carbon dioxide + a little energy

Yeast, a small single-celled fungus, can respire both aerobically and anaerobically. Its anaerobic respiration is often called fermentation, a process on which production of beer, wines, and spirits depends.

This is an investigation of fermentation of glucose by yeast, and the effect of ethanol, an end-product of the reaction, on the rate of fermentation.

Preparation

Materials

- 10 cm^3 suspension of yeast cells
- 10 cm^3 glucose solution
- 5 cm^3 ethanol
- Two 100 cm^3 beakers
- Two 10 cm^3 plastic syringes
- Two 30 cm lengths of capillary tubing
- Two short lengths of rubber tubing
- Retort stand, boss, and clamp
- Clock, or watch
- Ruler, graduated in millimetres
- Glass-marking pen

Method

1 Use one of the syringes to put the following into a beaker:
   5 cm^3 suspension of yeast cells
   5 cm^3 glucose solution
   5 cm^3 water

Gently stir the mixture, then fill the syringe to the 10 cm^3 mark. Fit a small piece of rubber tubing to the nozzle, and attach the capillary tube, as shown in Figure 43. Number the apparatus (syringe 1) and support it in the clamp.

Fig. 43 *Apparatus for measuring the rate of fermentation of sugar by yeast*

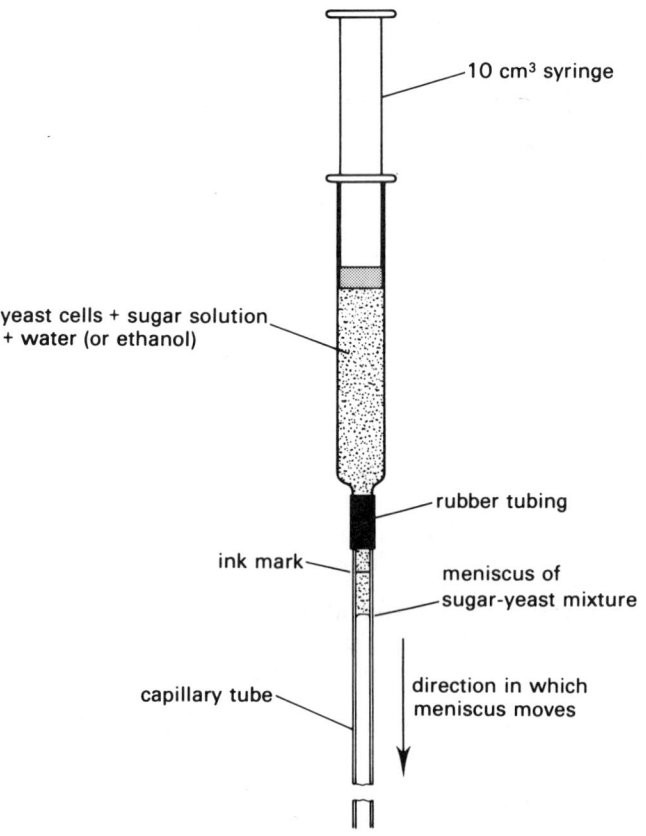

2 Use a clean syringe to put the following into a clean beaker:
   5 cm^3 suspension of yeast cells
   5 cm^3 glucose solution
   5 cm^3 ethanol

Stir the mixture and fill the syringe to the 10 cm³ mark. Fit a glass capillary tube by means of rubber tubing. Number the apparatus (syringe 2) and support it in the clamp.

3 Use the pen to make a mark just below the end of the rubber tubing on each piece of apparatus. Gently press the handle of each syringe to bring the meniscus up to the level of this mark. See Figure 43.

Observation

1 At intervals of 10 minutes, over a period of 70 minutes, record the distance between the meniscus and ink mark on each capillary tube. Tabulate your results.

Investigation

Materials

- Record of results
- Graph paper

Method

1 **Make a neat copy of the table showing your results.** (4)

2 **Draw a graph of your results, to compare rates of respiration in the two syringes.** (8)

3
(a) **In which syringe was respiration most rapid?** (1)

(b) **What was the effect of ethanol on the rate of respiration?** (1)

4 **Was respiration of the yeast cells mostly aerobic or anaerobic? Give a reason for your answer.** (2)

5 Syringe 2 contained a mixture of 5 cm³ suspension of yeast cells, 5 cm³ glucose solution and 5 cm³ ethanol. **What is the percentage of ethanol in the mixture? Show how you arrive at your answer.** (2)

6 **Explain how the apparatus works. Why does the meniscus move along the capillary tube?** (2)

Total 20 marks

Taking it further

1 Design an experiment to find out if yeast ferments sucrose more rapidly than glucose.
   (a) List the materials and apparatus for your experiment.
   (b) Write out your method, listing instructions in the order they should be carried out.
   (c) Draw a table, with headings, for your results.

Yeast cells budding

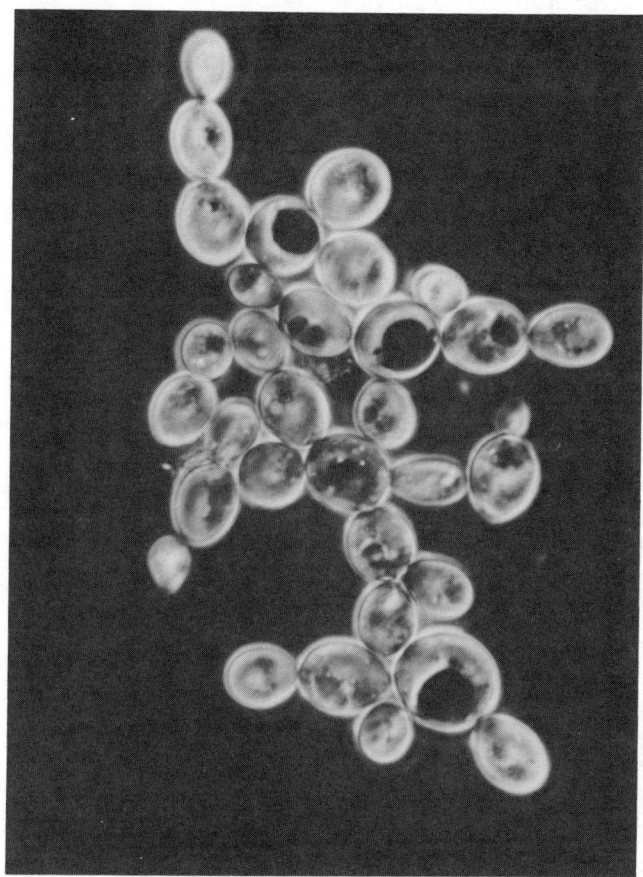

37 Making use of yeast

TIME
Preparation: 10–15 minutes
Observation: 7 days
Investigation: 60–80 minutes

Yeast is a saprophytic fungus found on the skins of fruit and surfaces of leaves. For many centuries it has been used in baking and brewing. Enzymes released by the yeast cells ferment sugars, producing ethanol (alcohol) and carbon dioxide:

glucose or sucrose $\xrightarrow{\text{enzymes}}$ ethanol + carbon dioxide

Other enzymes from the yeast break down proteins into amino acids. In baking, yeast is mixed with a thick sticky dough. Bubbles of carbon dioxide trapped in the dough cause the mixture to rise. Rising continues until a thin skin, formed chiefly from protein, is broken by enzymes and punctured by gas bubbles. At this point the dough begins to collapse.

In brewing, a solution of sugar is fermented to produce ethanol. Carbon dioxide, the other end product, may dissolve in the solution, forming small bubbles when a beer or wine shaken.

This investigation shows how yeast is used in baking and brewing, and gives ways of measuring the rate at which (a) dough rises; and (b) ethanol is formed from sugar.

Preparation

Materials

- 5 g dried yeast
- 150 cm³ sucrose solution
- 100 cm³ measuring cylinder
- Winemaker's plastic hydrometer
- Jam jar, or similar container
- Cotton wool

Method

1 Pour 100 cm³ of the sucrose solution into the measuring cylinder. Put the hydrometer into the cylinder, and measure the specific gravity of the solution, as shown in Figure 44. Write down your result.
2 Pour all of the sucrose solution into the jam jar, add the dried yeast, stir, and plug the neck of the jar with cotton wool. Stand the jar in a warm place.

Fig. 44 *Using a wine hydrometer to measure specific gravity (S.G.)*

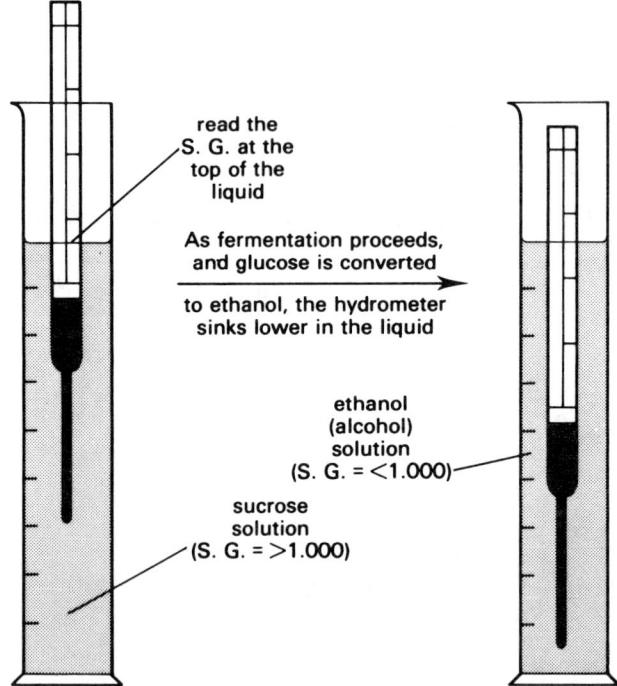

Observation

1 Each day, over a period of 7 days, pour 100 cm³ of the mixture into the measuring cylinder and record its specific gravity.

Fermentation is complete when the specific gravity falls below 1.000 and shows no further change.

Investigation

Materials

- 50 cm³ dough in a beaker
- 100 cm³ measuring cylinder
- 1 dm³ beaker
- Thermometer
- Bunsen burner, tripod and gauze
- Clock, or watch
- Record of results
- Graph paper

Method

Part A

1 Pour about 500 cm³ tap water into the 1 dm³ beaker. Place the beaker over a tripod and gauze, light the bunsen burner, and raise the temperature of the water to 35 °C. Maintain the water at this temperature.

2 Carefully pour dough into the measuring cylinder until it reaches the 30 cm³ mark, then stand the measuring cylinder inside the beaker, as shown in Figure 45. At intervals of five minutes, over a period of 45 minutes, record the volume of dough in the cylinder. **Tabulate your results.** (2)

While you are waiting for results, continue at question 5.

3 **Draw a graph of your results, showing the point at which the dough began to collapse.** (4)

4

(a) **Describe the shape of the graph.** (1)

(b) **What is the percentage of its orginal volume by which the dough has risen? Show how you arrive at your answer.** (2)

(c) **Why does the dough stop rising?** (2)

Part B

5 Refer to your results from the experiment with sucrose solution. **Make a neat copy of the table, showing daily changes in the specific gravity of the sucrose solution.** (4)

6 **How long did it take for all the sugar to be converted into ethanol?** (1)

7 Many people make beer and wine at home. **Suggest reasons for the following advice given to home brewers:**

(a) **Clean and sterilise all equipment before it is used.** (2)

(b) **Do not make beer or wine in glass vessels with screw tops. It is best to use a plastic container, fitted with an air lock.** (2)

Total 20 marks

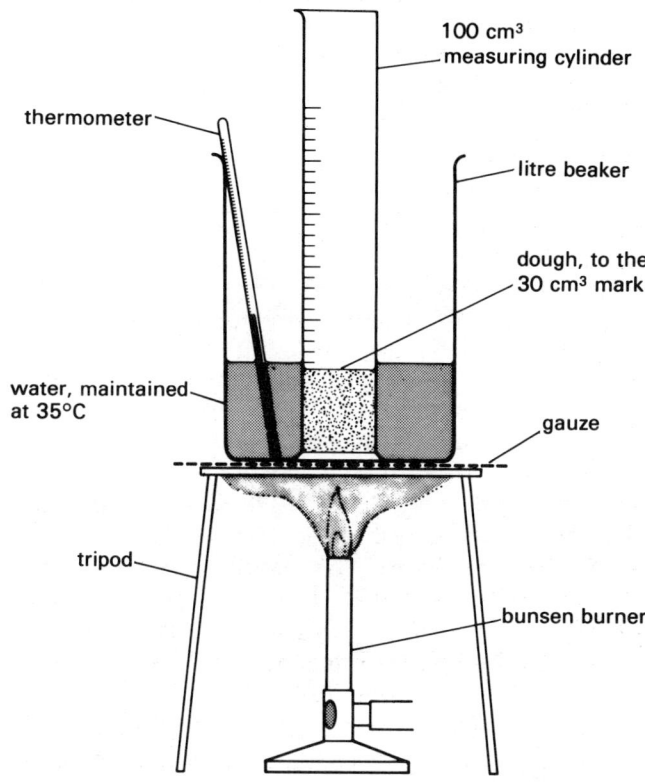

Fig. 45 *Apparatus for measuring the expansion of dough*

Taking it further

1 Design an experiment to find out if dough made from white flour rises faster than dough made from brown flour.
  (a) List the materials and apparatus for your experiment.
  (b) Write out your method, listing instructions in the order they should be carried out.
  (c) Draw a table, with headings, for your results.

2 It is claimed that adding caster sugar to dough makes the dough rise faster. How would you test this hypothesis?
  (a) State your hypothesis.
  (b) How would you test your hypothesis?
  (c) What do you predict would happen if your hypothesis was correct?
  (d) Draw a table, with headings, for your results.

38 The effects of a selective weedkiller on a lawn

TIME
Preparation: 30 minutes, and 30 minutes after 2–3 weeks
Investigation: 30–40 minutes
April–September

Fig. 46 *Plants of lawns*

(i) Side view

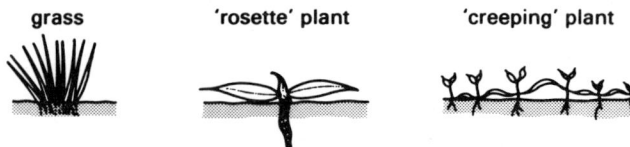

(ii) View from above

Fig. 47 *Estimating cover-values using a quadrat*

The 'rosette' plant has a cover-value of 19% because one or more leaves of the plant occur in 19 squares. Similarly, the 'creeping' plant has a cover-value of 8%

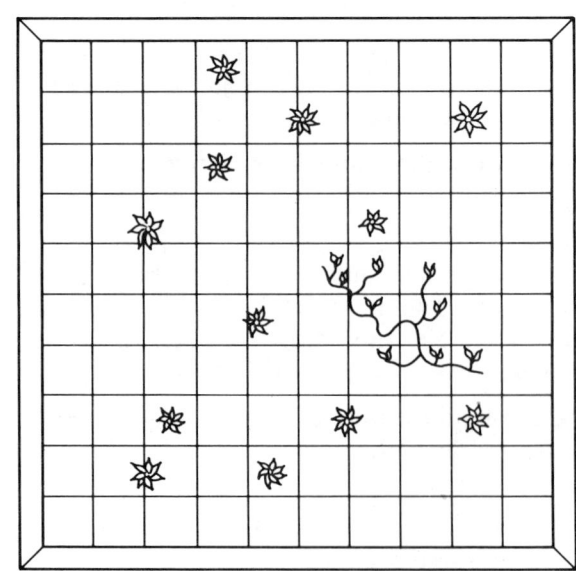

Chemical weedkillers, or herbicides, are widely used to control weeds in gardens and on farms. Total weedkillers kill all the weeds on paths and on land that needs to be cleared of vegetation. Selective weedkillers are used on lawns and on fields of cereals, such as wheat and barley. Most of the selective weedkillers used on lawns contain an auxin-like substance called 2, 4–D. Grasses are not affected, but many broad-leaved weeds are killed. Soon after treatment, the leaves of broad-leaved plants turn yellow and start to curl. Stems and roots grow faster than usual, with some twisting. Plants usually die 2–3 weeks after the weedkiller has been applied.

The aim of this investigation is to find out the effect of a selective weedkiller on three different types of plant: a grass, a 'rosette' plant, and a 'creeping' plant. See Figure 46. The numbers of each plant are counted using a frame called a quadrat. Most quadrats, like the one shown in Figure 47 are metre squares, with 100 sub-divisions. Two to three weeks after applying the weedkiller, recounts are made to find out if it has had any effect on the plants.

Preparation

Materials

- Selective weedkiller
- Metre quadrat
- Watering can, fitted with a fine rose
- Hammer
- Wooden pegs
- Illustrated guide to the British Flora
- Metre quadrat

Method

1. Find a lawn containing grass, a 'rosette' plant, and a 'creeping' plant. Use the flora to identify the plants you have chosen. Put the quadrat over an area of lawn containing several plants of each type. Count the cover-values of each type of plant and record your results. Tabulate your results.
2. Before lifting the quadrat put a wooden peg at each corner, and hammer it into the ground. Lift the quadrat, then ask your teacher to water the area with weedkiller.
3. After 2–3 weeks return to the same piece of lawn and apply the quadrat to its original position. Count the cover-values of each type of plant. Tabulate your results.

Investigation

Materials

- Record of results

Method

1. Copy Table 15 and enter your results. (*6*)

Table 15

Condition of lawn	Cover values		
	Grass	'Rosette' plant	'Creeping' plant
Before applying weedkiller			
After applying weedkiller			

2. **What percentage of each type of plant was killed by the weedkiller?** (*2*)
   (a) **Grass** (*1*)
   (b) **The 'rosette' plant** (*1*)
   (c) **The 'creeping' plant** (*1*)
   Show how you arrive at your answers.
3. **Against which type of plant is the weedkiller:**
   (a) **most effective;** (*2*)
   (b) **least effective?** (*1*)

4. **Why should selective weedkillers not be used in flower and vegetable gardens?** (*2*)
5. Mowings from lawns recently treated with selective weedkillers should be burned, not added to compost heaps. **Suggest a reason for this.** (*2*)
6. **Suggest ways in which the design of the experiment could have been improved.** (*2*)

Total 20 marks

Taking it further

1. The height at which a lawn is mown affects the types of weed that grow on it. Setting mower blades close to the soil surface encourages the growth of 'rosette' plants. Design an experiment to find out how the number of 'rosette' plants in a lawn is affected by mowing (i) at 1.5 cm, and (ii) 10 cm above the soil surface.
   (a) List the materials and apparatus for your experiment.
   (b) Write out your method, listing instructions in the order they should be carried out.
   (c) Draw a table, with headings, for your results.
2. When paths are made across a lawn, some plants are killed by trampling. Design an experiment to find out how trampling affects four different species of plant, including one grass.
   (a) List the materials and apparatus for your experiment.
   (b) Write out your method, listing instructions in the order they should be carried out.
   (c) Draw a table, with headings, for your results.

39 The effects of artificial fertiliser on a population of water plants

TIME
Preparation: 5–15 minutes
Observation: 6–10 weeks
Investigation: 40–50 minutes
May–September

A population is a group of organisms of the same species living in a particular place at the same time. The number of individuals in a population may increase, decrease, or remain constant. Numbers tend to increase if more food and space become available. Alternatively, numbers may fall if food supplies run short, or if individuals become overcrowded.

The subjects of this investigation are populations of duckweed plants, grown in beakers of water. As each population increases you should get an S-shaped growth curve, similar to that in Figure 48.

Fig. 48 *Typical population growth curve*

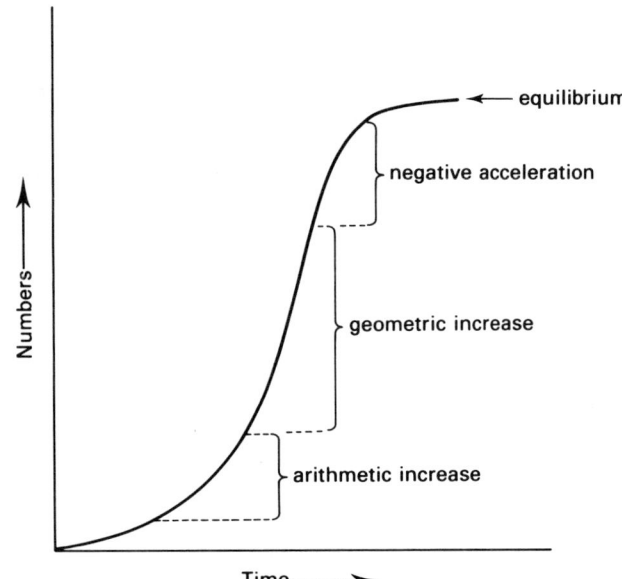

Growth curves are often subdivided into a number of phases of growth. The most important of these are as follows:

(i) Arithmetic increase: numbers increase in the succession 1, 2, 3, 4, 5, 6, etc.
(ii) Geometric increase: numbers increase in the succession 1, 2, 4, 8, 16, 32, etc.
(iii) Negative acceleration: the growth rate slows, as a result of a shortage of food, and the accumulation of waste products.
(iv) Equilibrium: numbers remain constant, as new individuals are born at the same rate that old individuals die.

Your aim is to find the effect of adding artificial fertiliser, a common pollutant of pond and river water, on the growth rate of a population of duckweed plants.

Preparation

Materials

- Duckweed plants
- Two 250 cm³ beakers
- Artificial fertiliser
- Glass–marking pen

Method

1 Fill the two 250 cm³ beakers with tap water, and put four plants of duckweed to each beaker. Label one beaker "artificial fertiliser" and add 2–3 granules of the fertiliser. Mark the water level in each beaker. Stand the beakers on a bench in the greenhouse, or tie them in polythene bags and put them in a suitable position out of doors. See Figure 49.

Observation

1 Each week, over a period of 6–10 weeks, count and record the number of plants in each beaker. If possible, make a count on the same day of each week. Tabulate your results.

Fig. 49 *Setting up Investigation 39*

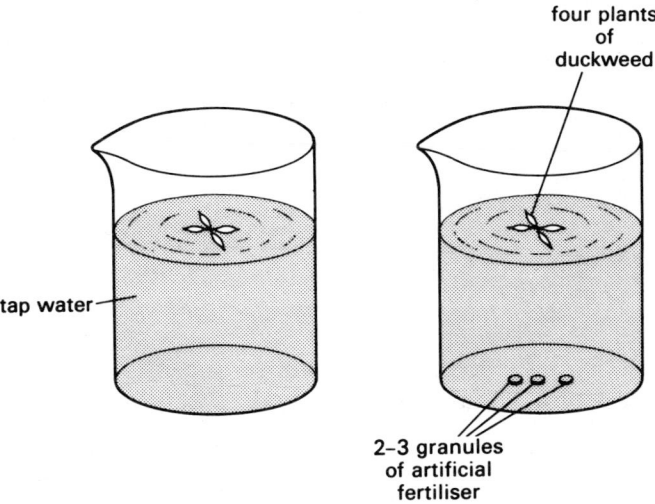

2 Add more tap water to each beaker if the water level falls.

Investigation

Materials

- Two populations of duckweed plants, in 250 cm³ beakers
- Individual records of population growth
- Graph paper

Method

1 **Make a neat copy of the table showing your results.** *(3)*

2 **Draw a graph of your results.** *(5)*

3 **Why should you count the number of plants on the same day of each week?** *(2)*

4

(a) **What was the effect of the fertiliser on the rate of population growth?** *(1)*

(b) **Suggest a reason for the fertiliser.** *(1)*

5 **Suggest a reason for the rate of growth in each beaker eventually slowing down.** *(1)*

6

(a) **Describe any difficulty you had in counting numbers of duckweed plants.** *(1)*

(b) **How could this difficulty be overcome?** *(1)*

7 Carefully examine the water in each beaker. **Comment on any differences between the two water samples.** *(2)*

8 **How might adding artificial fertiliser to pond water affect animal life in the pond?** *(3)*

Total 20 marks

Taking it further

1 It is claimed that adding artificial fertiliser to a pond reduces the number of water fleas. Design an experiment to test this hypothesis.
  (a) State your hypothesis.
  (b) How would you test your hypothesis?
  (c) What do you predict would happen if your hypothesis was correct?
  (d) Draw a table, with headings, for your results.

2 How would you investigate the effect of artificial fertiliser on the growth of microscopic animals in pond water?
  (a) List the materials and apparatus for your experiment.
  (b) Write out your method, listing instructions in the order they should be carried out.
  (c) Draw a table, with headings, for your results.

40 Some characteristics of a population of trees

TIME
Preparation: 40–60 minutes
Investigation: 30–40 minutes

Woodlands contain populations of trees. These may belong to the same species, or more typically to several different species. Each population usually consists of trees of different ages, some young, some middle-aged and some old. Examining and recording the age of individual trees is difficult, because it involves felling and counting the number of growth rings at the base of the trunk. A simpler method is to record size composition, found by measuring the circumference (girth) of tree trunks at breast height. Trees are then grouped according to the circumference of their trunks, and a histogram plotted, similar to that shown in Figure 50.

Fig.50 *Size distribution of oak trees in woodland*

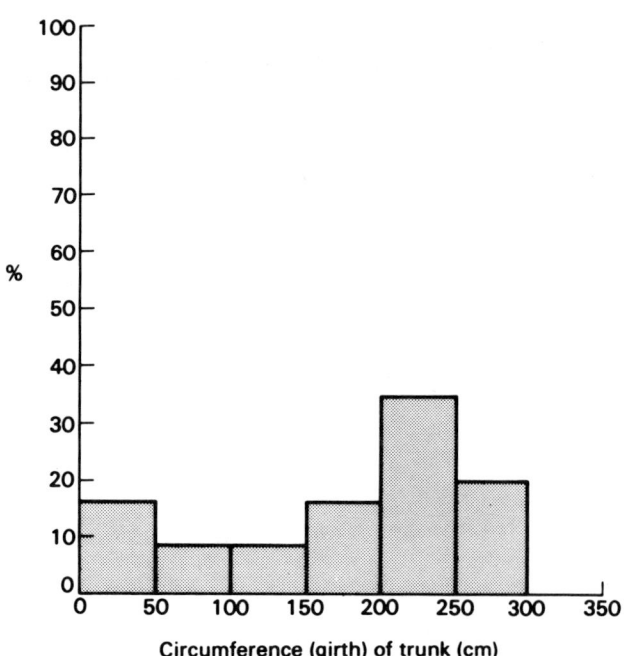

The aim of this investigation is to record the size composition of oak or beech trees in woodland. Additionally, you are asked to record if the trunk is colonised by ivy, a rooted climber that uses tree trunks for support.

Preparation

Materials

- 5 m tape measure, graduated in centimetres
- Copy of Table 16, mounted on cardboard
- Pencil

Method

1. Prepare a copy of the Table 16, and stick it to a piece of stiff cardboard. You will require the table, and a pencil, when recording the circumference of trees.
2. When you arrive at the wood, your teacher will help with the identification of oak, beech, and ivy. Working in pairs, choose either oak or beech. Measure the circumference of the nearest tree, and make your first entry in the table.
3. Moving outwards from the first tree, record the circumferences of the nearest 24 oak or beech trees. Additionally, immediately after recording the circumference of each tree, record if it is colonised by ivy. Take care not to omit any trees from within your sample area.

Investigation

Materials

- Table of results
- Graph paper

Method

Part A

1. **From your results plot a histogram, similar to Figure 50, showing the size distribution of oak or beech trees.** (6)
2. **Suggest a hypothesis to account for the size distribution shown in your histogram.** (4)
3. **Criticise the instruction to measure the trunks of trees at 'breast height'.** (2)

Table 16

Tree No.	Circumference of trunk (cm)	Colonisation by ivy (✓ or ×)
1		
2		
3		
4		
5		
6		
7		
8		
9		
10		
11		
12		
13		
14		
15		
16		
17		
18		
19		
20		
21		
22		
23		
24		
25		

Part B

4 In each column of your histogram shade the percentage of trees that are colonised by ivy. *(4)*

5 What conclusions can be drawn about the size of a tree trunk and its colonisation by ivy? *(2)*

6 Suggest a hypothesis to account for your conclusion. *(2)*

Total 20 marks

Taking it further

1 In a mixed oak/beech wood it is claimed that ivy is more commonly found on mature oaks than beeches. Design an experiment to test this hypothesis.
   (a) State your hypothesis.
   (b) How would you test your hypothesis.
   (c) What do you predict would happen if your hypothesis was correct?
   (d) Draw a table, with headings, for your results.

2 In a wood some beech trees grow in full sunlight, while others grow in shade. How would you find out if the leaves of trees growing in shade were larger than those growing in sunlight?
   (a) State your hypothesis.
   (b) How would you test your hypothesis?
   (c) What do you predict would happen if your hypothesis was correct?
   (d) Draw a table, with headings, for your results.
   (e) Suppose you find that shaded beech trees have larger leaves. How might the trees benefit from this condition?

41 Finding the size of an animal population

TIME
Preparation: 30 minutes, and 30 minutes after 4–5 days
Investigation: 30–40 minutes
April–September

Biologists may wish to find the size of an animal population, such as the number of snails in a garden or wood. Before this can be done, individual animals must be captured, marked, and released without being injured. The simplest formula for predicting the size of a population is called the Lincoln index. It is expressed by the following equation:

$$N = \frac{F_1 \times F_2}{F_3}$$

in which N is the estimated total population, F_1 is the number of marked individuals released into population, F_2 is the number of individuals captured in the second sample and F_3 is the number of marked individuals in the second sample. Firstly, you will use the formula in an attempt to find the size of a population of snails. Then, in order to get a better understanding of how the equation is used, you will carry out a laboratory exercise with marbles or beads.

Preparation

Materials

- Slug pellets
- White cellulose paint
- Paint brush

Method

1 Choose a small area of a garden or wasteland, preferably surrounded by wide paths, walls, or some physical barrier. Collect snails by turning stones and searching amongst vegetation. Mark the shell of each snail with a conspicuous white cross, using cellulose paint. Release the marked individuals into the population, putting them back in exactly the same positions in which they were found. Record the number of marked individuals (F_1).

2 Four days after marking the snails and returning them to the population, make another search and collect all of the snails that you can find. Alternatively, sprinkle the area with slug pellets, and return the following morning to count the number of snails killed. Record the number of snails that were captured in the second sample, or killed by slug pellets (F_2), and the number of these that were marked (F_3).

Investigation

Materials

- 50 marbles, or beads
- White cellulose paint
- Paint brush
- Cocoa tin, or opaque container with a lid
- Table spoon
- Petri dish base
- Record of results

Method

Part A

1 Copy Table 17 and enter your results. (3)

Table 17

Number of snails marked (F_1)	
Number of snails collected, or killed, in the second sample (F_2)	
Number of marked snails in the sample (F_3)	

2 **Use the Lincoln index to estimate the total size of the snail population**

$$N = \frac{F_1 \times F_2}{F_3}$$

(2)

3 **What percentage of the snail population was collected in the second sample, or killed by slug pellets? Show how you arrive at your answer.** (2)

4 **What percentage of the snail population was not captured in the second sample, or killed by slug pellets? Show how you arrive at your answer.** (2)

Part B

5 Mark 20 of the marbles with a small cross of white paint (F_1). Put all of the marked and unmarked marbles into the cocoa tin. You now have a "population" of 50 individuals, of which 20 (= 40%) are marked. Put the lid on the tin. Gently shake the marbles until marked individuals are evenly distributed throughout the "population". Remove the lid of the tin. Without looking into the tin, use the spoon to remove 10 marbles (F_2). Count and record the number of marked "individuals" in the sample (F_3). See Figure 51. **Use the Lincoln index to find out the approximate size of the marble "population". Show how you arrive at your answer.** (2)

Fig. 51 *'Recapturing' marked and unmarked individuals*

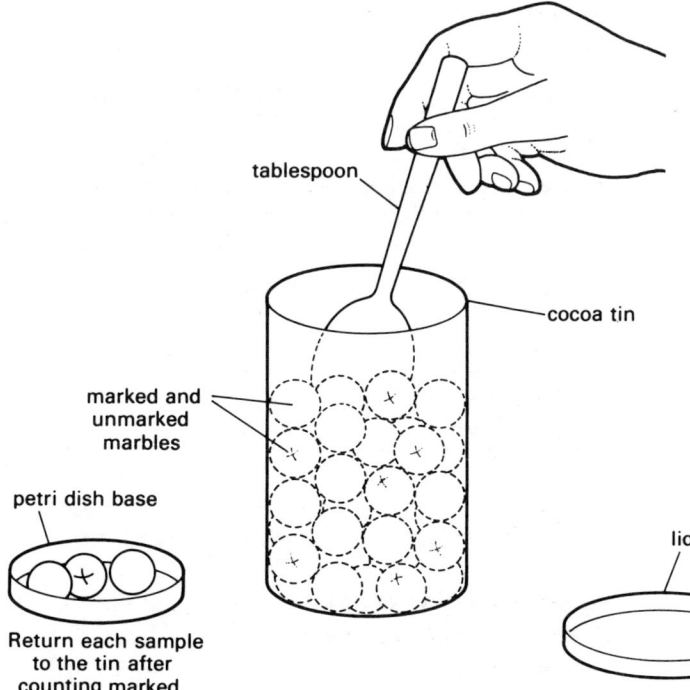

6 **Make two "recaptures" of 10 marbles. Estimate the size of the population after each "recapture". Show how you arrive at your answers.** (4)

7 You have made three different estimates of population size. **What is their mean (average) value?** (2)

8 There were 50 marbles in the tin. **What is the difference between the actual size of the population and the mean value you have obtained? Show how you arrive at your answer.** (1)

9 **Comment on the accuracy of the Lincoln index when it is used to estimate the size of a population.** (2)

Total 20 marks

Taking it further

1 How would you find out the percentage of a snail population killed by a single application of slug pellets?
   (a) List the materials and apparatus for your experiment.
   (b) Write out your method, listing instructions in the order they should be carried out.
   (c) Draw a table, with headings, for your results.
2 Snails feed on a wide variety of garden vegetables. Even so, they prefer some vegetables to others. If one or more snails is put into a plastic box, with a selection of vegetables, it is possible to find out which ones they prefer. You will, of course, need to measure the mass of food eaten. Design an experiment to find out the food preference of a garden snail.
   (a) List the materials and apparatus for your experiment.
   (b) Write out your method, listing instructions in the order they should be carried out.
   (c) Draw a table, with headings, for your results.
   (d) Comment on possible sources of error.

42 Pollution of water and air

TIME
Preparation: 10–15 minutes and 15–20 minutes
Observation: 4 weeks
Investigation: 30–40 minutes
April–September

Pure water and air probably don't exist in nature. Even so, the water we drink, and the air we breathe, must be clean if the human population is to enjoy good health. Water or air that contain substances harmful to our health, or the health of other organisms, is said to be polluted. Water may be polluted by fertilisers, or insecticides, washed from fields, or chemicals released from factories.

Finding the extent to which water and air samples are polluted is often very difficult. If you expect to find large amounts of any polluting substance, you will probably be disappointed. Amounts of polluting substances are often so small that the only way of detecting them is by measuring their effects on living organisms. This investigation involves two such tests:

(a) Duckweed, a small plant found floating in ponds, is very sensitive to the presence of pollutants in water. In the presence of sewage or fertilisers, its growth is increased. When insecticides or other poisonous chemicals are present, its growth rate is reduced.
(b) A large number of microscopic fungi grow on the underside of leaves. If air is polluted, the number of these fungi is reduced.

Preparation

Materials

Part 1
- Plants of duckweed
- Three 1 dm^3 beakers, or Kilner jars
- Glass-marking pen

Part 2
- Leaves of a known tree
- Nutrient, or malt agar plate
- Vaseline
- Cork-borer (No. 7–10)
- Incubator, kept at 35°C
- Glass-marking pen

Method

1. Four weeks before the assessment, label three 1 dm^3 beakers, or Kilner jars, A, B and C. Fill beaker A with distilled water, beaker B with tap water, and beaker C with pond or river water. Put four duckweed plants into each beaker. Stand the beakers on a bench in a greenhouse, or tie each beaker in a polythene bag and place it in a suitable position out of doors.

2. Two days before the assessment, select a known tree for your study of air pollution. Collect leaves from that tree in a place where you suspect the air may be polluted, such as a roadside verge, or around a factory. Collect leaves of the same tree from a different site, where there is no obvious source of pollution.

3. Use the cork-borer to cut three discs from a leaf taken from each site. Smear vaseline on the upper epidermis of each disc, and use forceps to stick it inside the lid of the petri dish, as shown in Figure 52. Allow the dish to stand on the bench surface for 15 minutes, gently tap it several times, then turn it upside down and incubate it at 35°C for 24 hours.

Observation

1. Each week, over a period of four weeks, count and record the number of duckweed plants in each beaker.
2. After incubating the plates, count the number of fungal colonies beneath each disc. These may be yellow, brown or grey in colour, either smooth or gelatinous or made up of five branching threads called hyphae. Tabulate your results.

Fig. 52 *Setting up Investigation 42*

(i) Duckweed in a beaker of water.

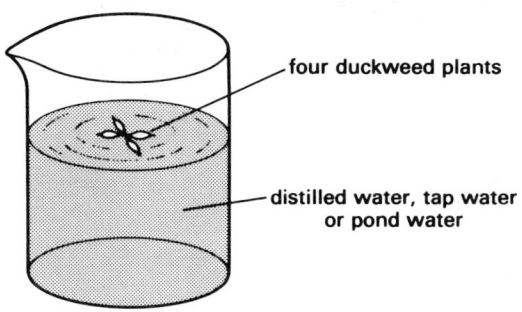

(ii) Leaf discs above agar

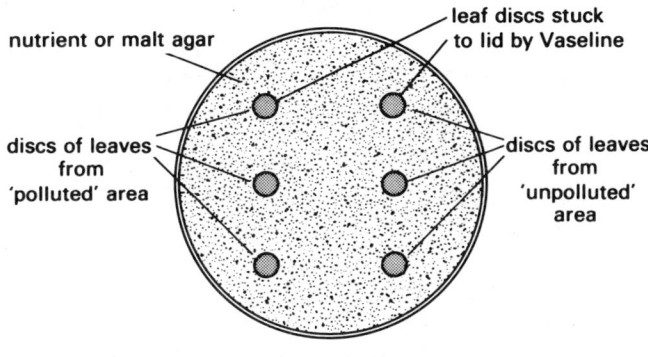

Fig. 53 *Form of the graph*

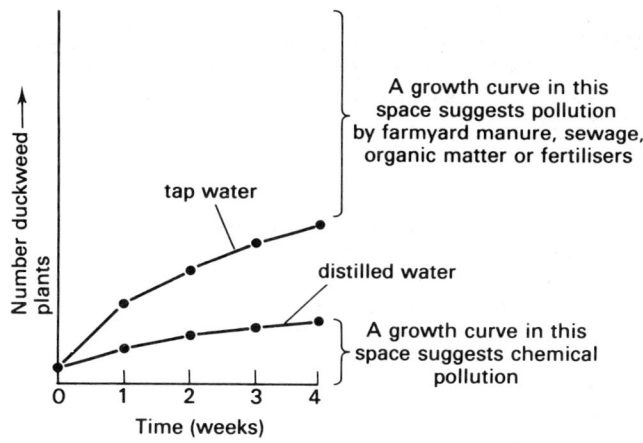

....
Investigation
....

Materials

- Record of results
- Graph paper

Method

Part A

1 Tabulate your results for the growth of duckweed plants in the following:

(a) **distilled water;** *(1)*

(b) **tap water;** *(1)*

(c) **pond or river water.** *(1)*

2 **Draw a graph of your results.** See Figure 53. (If the duckweed grows faster in pond or river water than in tap water, there is probably some pollution from soil, sewage, fertilisers or decaying organic matter. If growth is slower than in distilled water, some industrial pollutants may be present.) *(5)*

3 **What do you conclude from your investigation of the pond or river water?** *(2)*

Part B

4

(a) **Make a table to show the number of fungal colonies that grew beneath each leaf disc.** *(4)*

(b) **Calculate the mean (average) number of fungal colonies produced by leaf discs collected at each site. Show how you arrive at your answer.** *(2)*

(c) **What do you conclude from your results?** *(2)*

5 **Why was the dish allowed to stand on the bench for 15 minutes, tapped several times, then turned upside down before incubation?** *(2)*

Total 20 marks

....
Taking it further
....

1 Local people say that a river downstream from a factory is polluted by chemical wastes. Design an experiment to test this hypothesis.
 (a) State your hypothesis.
 (b) How would you test your hypothesis?
 (c) What do you predict would happen if your hypothesis was correct?
 (d) Draw a table, with headings, for your results.

2 Design an experiment to find out if there are more fungi on the underside of an oak or beech leaf.
 (a) List the materials and apparatus for your experiment.
 (b) Write out your method, listing instructions in the order they should be carried out.
 (c) Draw a table, with headings, for your results.

43 Organic pollution of water samples

TIME
Investigation: 40–50 minutes

Water in ponds, lakes and rivers is often polluted by organic material, such as leaf litter and dead animals. The presence of this material in the water causes an increase in the number of micro-organisms. Populations of bacteria and fungi are the first to increase, followed by populations of microscopic animals and algae. The more organic material in the water, the more micro-organisms it usually contains.

All of these micro-organisms produce an enzyme, catalase, that rapidly breaks down hydrogen peroxide into oxygen and water.

hydrogen peroxide ⟶ oxygen + water
(liquid)　　　　　　　(gas)　　(liquid)

The rate at which oxygen is released depends on the number of micro-organisms in the water. Water that is heavily polluted by organic material will therefore break down hydrogen peroxide solution more rapidly than unpolluted water. In this investigation you will mix equal volumes of water and hydrogen peroxide solution, and measure the rate at which oxygen is released. If bubbles of oxygen are trapped in the barrel of a syringe, an equivalent volume of the mixture is displaced into a capillary tube, where it can be measured.

Investigation

Materials

- 10 cm^3 pond water, in a beaker
- 10 cm^3 tap water, in a beaker
- 5 cm^3 5-vol. hydrogen peroxide solution
- Two 1 cm^3 plastic syringes
- Two 20 cm lengths of capillary tubing
- Two 2 cm lengths of rubber tubing
- Boss, clamp, and retort stand
- Stop-clock, or watch with a second hand
- Ruler, graduated in millimetres
- Glass-marking pen
- Safety spectacles
- Plastic gloves

Method

> **SAFETY PRECAUTIONS**
>
> *Wear safety spectacles and plastic gloves. Wash off any spillages with plenty of water.*

1 Put on your safety spectacles and plastic gloves. Number the syringes 1 and 2. Draw hydrogen peroxide solution into syringe 1 to the 0.5 cm^3 mark. Draw tap water into the same syringe until the mixture reaches the 1.0 cm^3 mark. Gently rock the contents of the syringe to assist mixing. Attach rubber tubing to the nozzle of the syringe, and fit a capillary tube, as shown in Figure 54. Support the apparatus in a clamp, at about 25 cm above the bench surface.

2 Similarly, set up syringe 2 with 0.5 cm^3 hydrogen peroxide solution and 0.5 cm^3 pond water.

3 Apply gentle pressure to the handle of each syringe until a meniscus appears at the top of each capillary tube. Mark the position of each meniscus.

4 At intervals of 1 minute, over a period of 5 minutes, mark the position of the meniscus in each capillary tube. Record your results.

5 **Tabulate your results.** (4)

6 **Draw a graph of your results.** (6)

7 **What conclusions can be drawn?** (3)

8 **Design an experiment to find out if water from the top or bottom of a pond is most heavily polluted by organic matter. Write out your method, listing instructions in the order they should be carried out.** (7)

Total 20 marks

Fig. 54 *Apparatus for Investigation 43.*

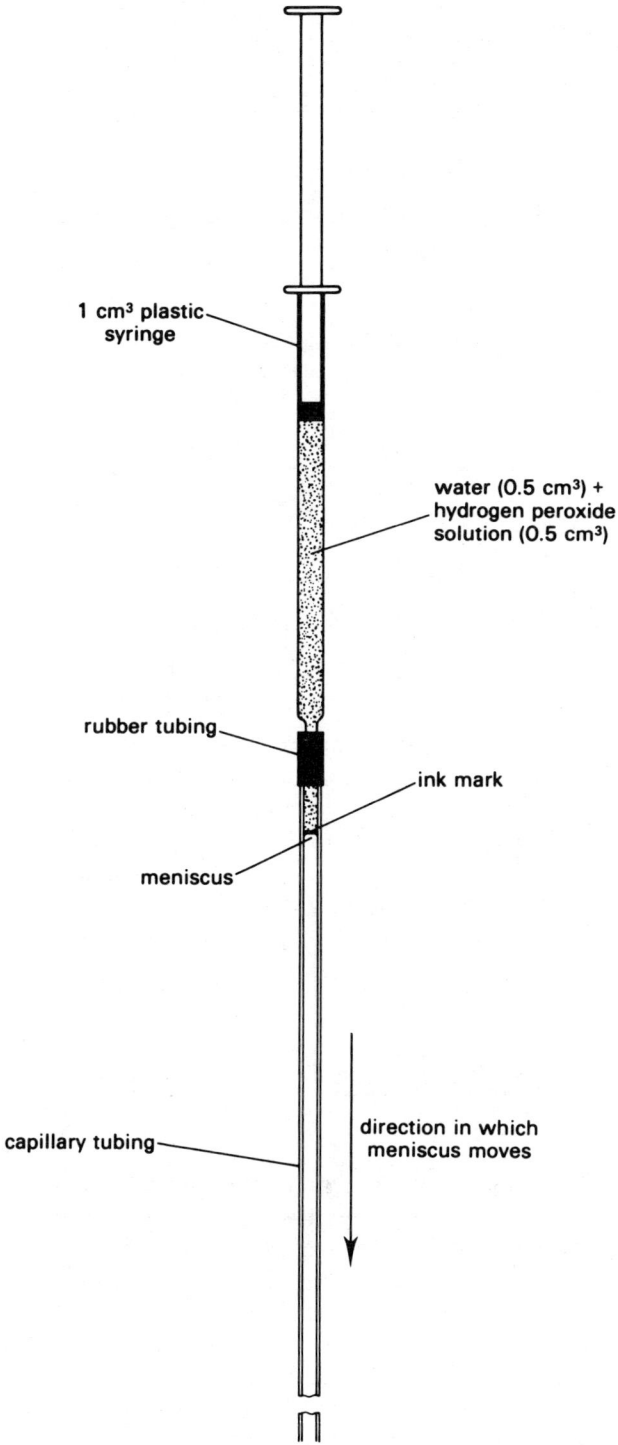

Taking it further

1 Artificial sea water can be prepared by dissolving tablets, containing mineral salts, in distilled water. Design an experiment to find out if a sample of sea water, collected from a beach, is polluted with organic matter.
   (a) List the materials and apparatus for your experiment.
   (b) Write out your method, listing instructions in the order they should be carried out.
   (c) Draw a table, with headings, for your results.

44 Water retention by sand, loam and peat soils

TIME
Investigation: 30–40 minutes

The amount of water in a soil affects the growth of plants. If cultivated soils are to produce good crops, they must be moist in summer and well-drained in winter. Each time it rains some water is held in the soil, mostly by fine particles of clay and organic matter. Surplus water drains away. Unfortunately, this drainage water may carry useful mineral salts to lower levels in the soil, out of reach of plant roots.

This is an investigation of the water-retaining properties of coarse sand, loam, and peat. Sand is a mineral soil, without any organic matter. Peat is organic, without any mineral particles. In loam there is a mixture of minerals and organic materials with particles of several different sizes.

Investigation

Materials

- Coarse sand, labelled **A**
- Loam, labelled **B**
- Peat, labelled **C**
- Three plastic flower pots
- Three 250 cm³ beakers
- 200 cm³ plastic beaker
- 100 cm³ plastic measuring cylinder
- Glass-marking pen

Method

Part A

1 Put a flower pot inside each 250 cm³ beaker, and number the pots from 1–3. Fill the 200 cm³ beaker up to the brim with coarse sand, level it off, and tip it into pot 1. Similarly, put the same amount of loam into pot 2, and peat into pot 3. Level off the soil in each pot.

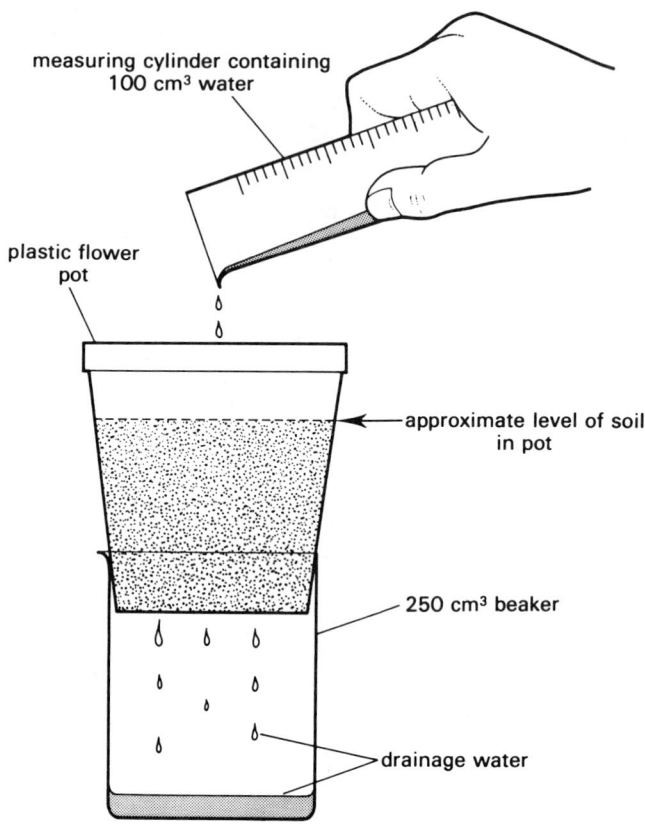

Fig. 55 *Apparatus for determining water retention by a soil*

2 Pour 100 cm³ tap water into the measuring cylinder, and gently pour it over the coarse sand, as shown in Figure 55.

Similarly, pour 100 cm³ tap water over (i) the loam soil and (ii) the peat soil. Wait until the drainage into the 250 cm³ beakers is complete. **Through which of the soils was drainage of water:**

(a) **most rapid;** (*1*)

(b) **slowest?** (*1*)

3 Take each 250 cm³ beaker in turn, and pour the water it contains into the measuring cylinder. Record the volume of water in each beaker.

(a) **Copy and complete Table 18.** (6)

(b) **What percentage of the water poured over loam soil has been retained? Show how you arrive at your answer.** (2)

(c) **Suggest ways in which the design of the experiment could have been improved.** (2)

Table 18

Soil	Sand (A)	Loam (B)	Peat (C)
Volume of water collected in beaker (cm³)			
Volume of water retained by soil (cm³)			

Part B

4 Empty the measuring cylinder and fill it to the 50 cm³ mark with loam soil, gently tapping it on the bench surface to prevent particles sticking to the sides. Pour in water until the mixture reaches the 100 cm³ mark. Place the palm of your hand over the open end of the measuring cylinder and gently shake the mixture up and down. After the soil and water have been thoroughly mixed, stand the cylinder on the bench and allow the particles to settle. Within a few minutes the soil particles in the measuring cylinder should have separated, as shown in Figure 56.

5

(a) **Copy and complete Table 19. What is the percentage of each type of soil particle in loam?** (4)

Table 19

Type of soil particle	% composition
Humus	
Clay	
Sand	
Gravel	

Fig. 56 *Separation of particles from a loam soil*

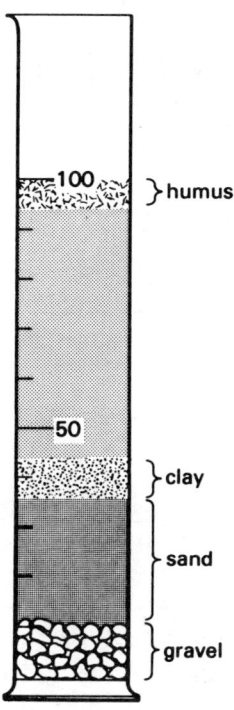

(b) **Show how you arrived at your answers.** (2)

(c) **Which particles in a loam soil play the greatest part in retaining water?** (2)

Total 20 marks

Taking it further

1 Keen gardners say that the deeper you dig in winter, the wetter the soil becomes. Design an experiment to test this hypothesis.
   (a) State your hypothesis.
   (b) How would you test your hypothesis?
   (c) What do you predict would happen if your hypothesis was correct?
   (d) Draw a table, with headings, for results.
2 Peat is added to soil to improve its water-retaining capacity. Suppose you dug a bale of peat into half your garden, thoroughly mixing it with the soil. How would you find out if the treated soil retained more water?
   (a) List the materials and apparatus for your experiment.
   (b) Write out your method, listing instructions in the order they should be carried out.
   (c) Draw a table, with headings, for your results.

45 Soil pH

TIME
Investigation: 60–90 minutes

The degree of acidity or alkalinity of a soil is expressed in terms of pH. The pH scale ranges from 1–14. Low values from pH 3.0–6.0 show that a soil is acid. High values from pH 8.0–10.0 show that a soil is alkaline. The ideal value for a garden soil is pH 6.5–7.0, which suits most flowers, fruits, and vegetables. Certain substances, such as peat or lime may be added to garden soil to change the pH to a suitable level. The addition of fertilisers and manures can also change pH.

The pH of any soil may be found by adding universal indicator, a mixture of dyes that changes colour as pH changes. See Figure 57. Your aim is to use universal indicator to find the pH of soil samples, and to measure changes in pH following the addition of urea, an organic fertiliser, to garden soil. Micro-organisms in the soil produce urease, an enzyme that breaks down urea into ammonium carbonate, an alkaline compound.

urea ⟶ ammonia + ⟶ ammonium
(neutral)   carbon dioxide   carbonate
                                 (alkaline)

Fig. 57

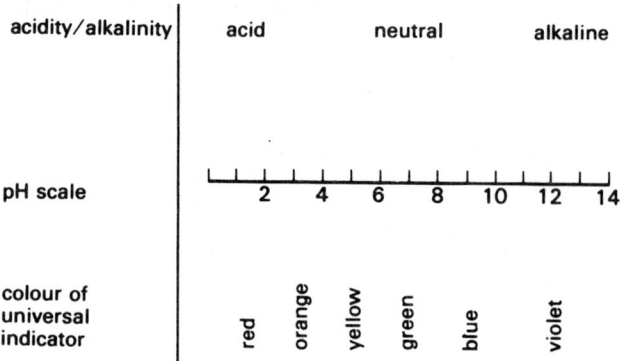

Investigation

Materials

- 50 g garden top-soil, in a 250 cm³ beaker
- 2 g peat
- 2 g loam
- 2 g lime
- 50 cm³ urea solution
- 25 cm³ distilled water
- Universal indicator
- Universal indicator chart
- Five test tubes
- Narrow range pH 6.0–8.0 papers
- Glass rod
- Spatula
- Forceps
- Glass-marking pen
- Graph paper

Method

1 Add the urea solution to the garden soil and stir the mixture with a glass rod. Use forceps to dip a small piece of indicator paper into the mixture. Record the pH. At intervals of 10 minutes, over a period of 60 minutes, dip pieces of indicator paper into the mixture and record any changes in pH. **Tabulate your results.** (2)

2 **Draw a graph of your results, to show how pH changed after urea was added to the soil.** (4)

3 Urea, produced by an industrial process, is sometimes used as a fertiliser. (1)

4 Draw three horizontal lines, approximately 1 cm apart, from the base of each test tube. When testing for pH, add the material to be tested to the first mark, distilled water to the second, and universal indicator to the third. Gently shake the mixture, and allow particles to settle before reading pH from the indicator chart. See Figure 58.

Fig. 58 *Measuring pH*

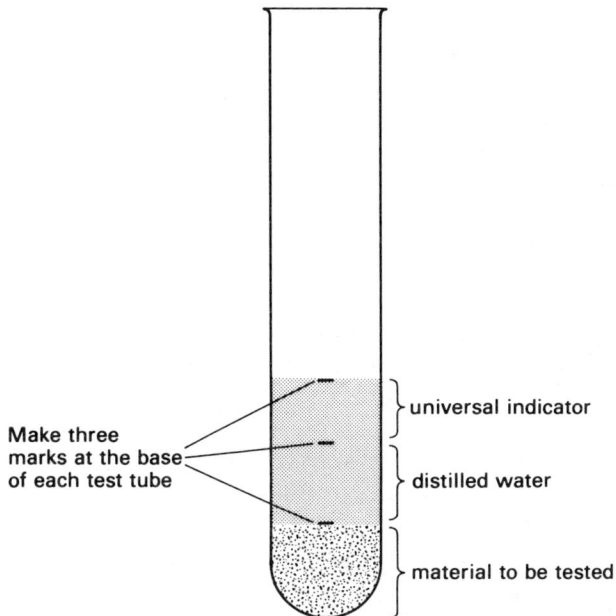

Fig. 59 *The effect of pH on the uptake of four elements by plant roots*

The thicker the band, the more of the element can enter roots.

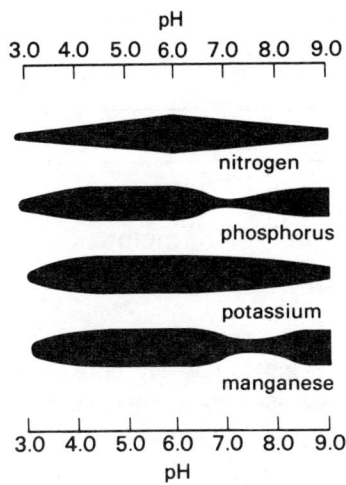

Measure and record the pH of each of the following:

(a) **Urea solution** *(1)*

(b) **Distilled water** *(1)*

(c) **Peat** *(1)*

(d) **Loam** *(1)*

(e) **Lime.** *(1)*

5 **Which of the substances listed above would you add to garden soil with the following pH values?**

(a) **5.6** *(1)*

(b) **7.5** *(1)*

6 **If peat and lime were mixed together, what effect would this have on the pH of the mixture?** *(2)*

7 Figure 59 shows the effect of pH on the uptake of mineral elements by plant roots. A wide band indicates that the element concerned is easily taken up at that particular pH. **Why is a neutral (pH 7.0) not ideal for the growth of most plants?** *(2)*

8 Urea is added to soil to supply nitrogen. If an excess of urea is added, plants may grow less well. **Explain the reasons for this.** *(2)*

Total 20 marks

Taking it further

1 Lime is added to garden soil to reduce acidity. However, the effect doesn't last for very long, because some of the lime is washed away by rain. Design an experiment to find out how adding lime to a soil affected pH (i) two weeks, and (ii) six months after its application.
   (a) List the materials and apparatus for your experiment.
   (b) Write out your method, listing instructions in the order they should be carried out.
   (c) Draw a table, with headings, for your results.

2 Design an experiment to find out how pH changes with increasing depth in a soil.
   (a) List the materials and apparatus for your experiment.
   (b) Write out your method, listing instructions in the order they should be carried out.
   (c) Draw a table, with headings, for your results.

46 Organic matter in soil

TIME
Preparation: 5–10 minutes
Investigation: 60–90 minutes

When plants and animals die their bodies are added to the soil as organic matter. This forms an important part of the soil. It retains water and increases the amount of air present. In addition, it serves as food for soil invertebrates, fungi and bacteria. Eventually, the micro-organisms break down organic matter into gases, water, and mineral salts. Micro-organisms in the soil produce many enzymes in order to break down organic matter. One of these enzymes, invertase, is responsible for the rapid breakdown of sucrose into simpler sugars.

$$\text{sucrose} \xrightarrow{\text{invertase}} \text{glucose} + \text{fructose}$$

Those fungi and bacteria responsible for breaking down leaf litter produce large amounts of invertase. Therefore, levels of invertase activity provide a useful guide to the amount of dead plant matter in a soil.

In this investigation you will use a physical and a chemical method to find out if the amount of organic matter in topsoil and subsoil collected from the same site. The physical method involves heating soil to burn off organic matter. In the chemical method, the soil is tested for the amount of invertase it contains. Which method on your view will provide the most accurate results?

Preparation

Materials

- Two polythene bags
- Spade, or garden trowel
- Pestle and mortar
- Incubator or oven, kept at 35°C
- Glass-marking pen

Method

1 Collect about 100 g topsoil from a vegetable garden, and 100 g subsoil from the same site. Put each sample into a polythene bag, label the bags, and take them back to the laboratory.
2 Dry both soil samples by heating them at 35°C for 24 hours. Grind the soils in a mortar until they are in the form of fine granules.

Investigation

Materials

- 70 g topsoil, 70 g subsoil
- 100 cm³ sucrose solution
- Two 100 cm³ beakers
- Eight Clinistix reagent strips
- Two glass rods
- Two metal sand trays
- Bunsen burner and tripod
- Spatula and tong
- Top-pan balance
- Aluminium foil
- Glass-marking pen
- Safety spectacles

Method

> **SAFETY PRECAUTIONS**
>
> *Wear safety spectacles when heating soil on a sand tray.*

1

(a) Weigh 50 g topsoil and 50 g subsoil. Put each sample into a beaker and label the beakers. Add 50 cm³ sucrose solution to each beaker, and stir the mixtures with a glass rod. Immediately, and at intervals of 15 minutes over a period of 45 minutes, dip a Clinistix reagent strip into each beaker. Use Table 1 (page 8) to find the amount of glucose in each mixture. **Tabulate your results.** While you are waiting for results, continue at question 2. (*2*)

(b) **What do you conclude from your results?** (*3*)

2

(a) Put on your safety spectacles. Weigh 20 g topsoil and spread it out on a sand tray. Stand the sand tray over a tripod and heat it with the bunsen burner for 5–10 minutes, turning the soil with a spatula. See Figure 60. Use the tongs to rest the sand tray on several thicknesses of aluminium foil. Allow the tray to cool.

Fig. 60 *Method of heating soil in a sand tray*

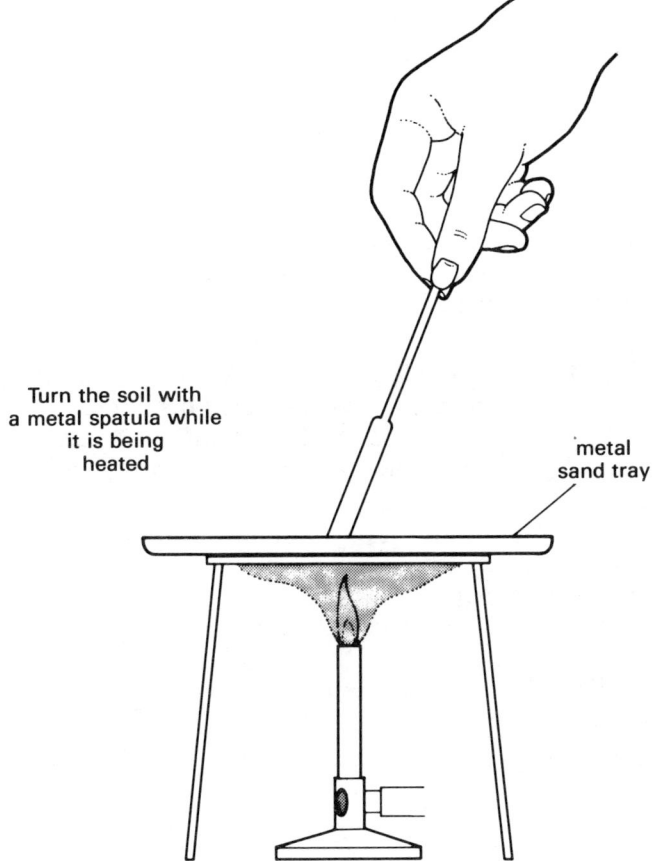

(b) Similarly, weigh 20 g subsoil and treat it in exactly the same way as the topsoil. After heating for 5–10 minutes, allow the tray to cool.

3 **During heating, what happened to:**

(a) **the organic matter;** (*1*)

(b) **the inorganic matter?** (*1*)

4 When the soils have cooled, weigh each soil sample and record its mass. **Copy and complete Table 20.** (*4*)

Table 20

	Topsoil	Subsoil
Mass before heating (g)	20	20
Mass after heating (g)		
Mass of organic matter (g)		

5 **Calculate the percentage of organic matter in:**

(a) **topsoil;** (*2*)

(b) **subsoil.** (*2*)

Show how you arrive at your answers.

6 **List three possible sources of error in the experiment.** (*3*)

7 A detective believes that soil on the boots of a murder victim came from a suspect's garden. **How would you test this hypothesis?** (*2*)

Total 20 marks

Taking it further

1 A gardner says his soil contains more organic matter in November than in June. Design an experiment to test this hypothesis.
 (a) State your hypothesis.
 (b) How would you test your hypothesis?
 (c) What do you predict would happen if your hypothesis was correct?
 (d) Draw a table, with headings, for your results.

47 Decomposers in the soil

TIME
Preparation: 10–15 minutes
Observation: 2–4 weeks
Investigation: 30–40 minutes

Decomposition is the breakdown of dead plants and animals. Dead organic matter, including leaf litter, is decomposed by a number of different organisms. Invertebrate animals, such as woodlice, earthworms, millipedes, and snails shred the fallen leaves into smaller particles. These smaller particles are then usually colonised by bacteria and fungi, feeding as saprophytes. The saprophytes complete the process of breakdown. As decomposition nears completion, carbon dioxide mineral salts, and water are released into the surrounding soil.

Rates at which decomposition occurs can be found by measuring the rate at which leaf discs are broken down, or removed from fixed positions on a lawn, or beneath the soil surface. In this investigation you will find out the fate of 25 leaf discs, pinned to the soil surface.

Preparation

Materials

- Leaves of oak or beech
- No. 14 cork-borer (diameter = 2.0 cm)
- 25 pins
- Marker cane

Method

1 Using the cork-borer, cut 25 discs from the leaves.
2 Find a secluded part of a lawn, garden, or floor of a wood.
3 Place a pin through each leaf disc, as illustrated in Figure 61. Arrange the discs in five rows on the soil surface, roughly 5 cm apart. Pin each to the soil. Mark the position of your discs with the marker.

Fig. 61 *Method of anchoring a leaf disc with a pin*

The pin should be placed close to the edge of the disc.

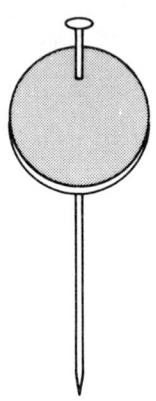

Observation

1 Each week, over a period of 4–8 weeks, count:
  (a) the number of discs that have been removed;
  (b) the number of discs that have been damaged by feeding animals. Keep a record of your results.

Investigation

Materials

- Record of results
- Graph paper

Method

Part A

1 **Draw a graph of your results.** (5)

2 **Suggest a reason for placing the pin near the edge of the leaf disc, and not through its centre.** (1)

3 **In addition to invertebrate animals, name three other factors that may have caused some leaf discs to be removed from their original position.** (3)

4 How could each of the following be determined?

(a) **The total area of the 25 leaf discs?** (Area of circle = πr^2, where $\pi = 3.14$) (3)

(b) **The mean (average) loss in area caused by animal feeding.** (3)

Part B

5 In another investigation 25 leaf discs, contained in three nylon bags with different sizes of mesh buried in soil to provide food for decomposers. The size of the mesh, and its effects on different decomposers, is shown in Figure 62.

Fig. 62 *Using nylon net bags to determine the effects of decomposers*

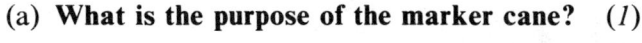

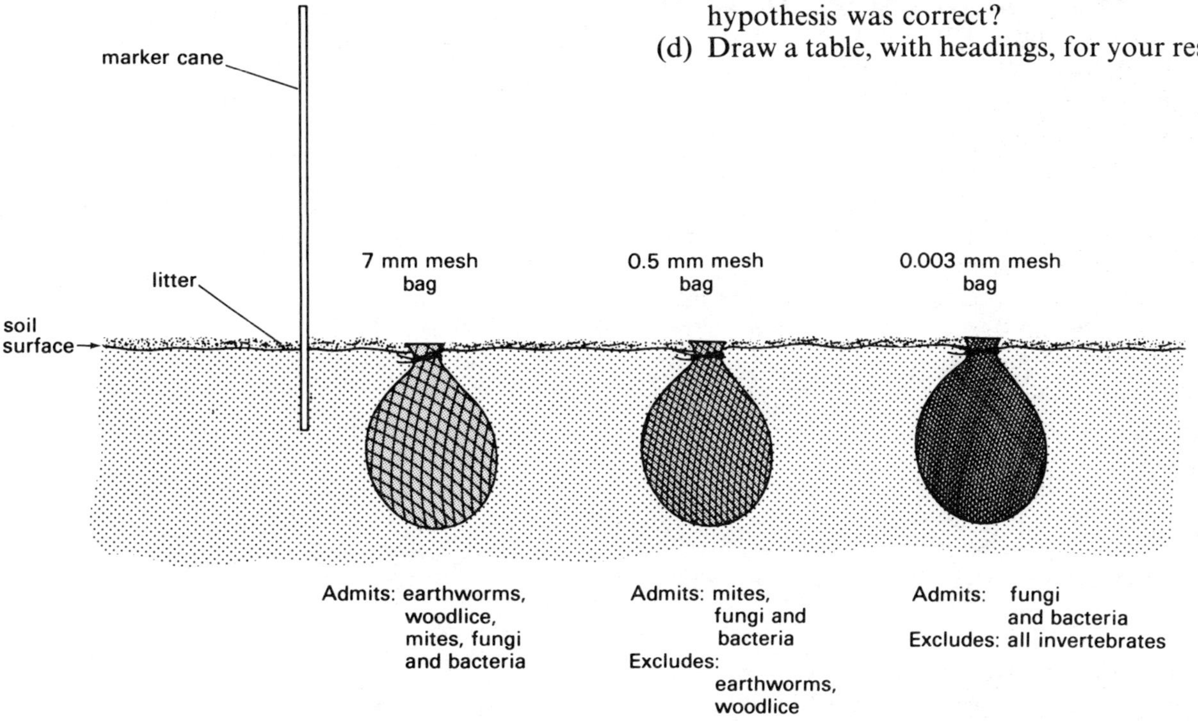

(a) **What is the purpose of the marker cane?** (*1*)

(b) **In which bag would you expect to find most rapid decomposition? Suggest a reason for your answer.** (*2*)

(c) **Suggest a suitable control for this investigation.** (*2*)

Total 20 marks

Taking it further

1 A scientist says that leaves tied in nylon tights decompose slower than those tied in a fibre sack. Design an experiment to test this hypothesis.
   (a) State your hypothesis.
   (b) How would you test your hypothesis?
   (c) What do you predict would happen if your hypothesis was correct?
   (d) Draw a table, with headings, for your results.

48 Gene recombination

TIME
Investigation: 30–40 minutes

The genes which determine a particular characteriestic occur in pairs. If both genes are similar (*AA* or *aa*) the individual is homozygous. If they are different (*Aa*) the individual is heterozygous. When sex cells are formed, the genes of each pair are separated. Sperms and eggs carry only one gene, either *A* or *a*. At fertilisation, the gametes (*A* or *a*) fuse in a random manner to produce a zygote with a pair of genes (*AA*, *Aa* or *aa*).

This investigation uses coloured marbles to represent genes. The aim is to make crosses between individuals of different genotypes, and to record the genotypes of their offspring. Red marbles represent the *A* gene and blue marbles the *a* gene.

Investigation

Materials

- Four red marbles
- Three blue marbles
- Two teaspoons
- Two opaque paper cups, tins, or similar containers

Fig. 63 *Removing a marble from the container. Each marble represents a gene contributed by one of the parents.*

Method

1 Make a cross between two individuals with the genotypes *AA* and *AA*. Put two red marbles (which represent the parents' genes) into each container. Each parent contributes one gene to an offspring. Without looking into the containers, use the spoon to remove one marble from each container.

(a) **What is the genotype of the offspring?** (*1*)

(b) **Is it possible for any of the offspring to have a different genotype?** (*1*)

(c) **What biological process is represented by taking one marble from each container and bringing them together?** (*1*)

(d) **Why is each parent represented by two marbles, not one?** (*1*)

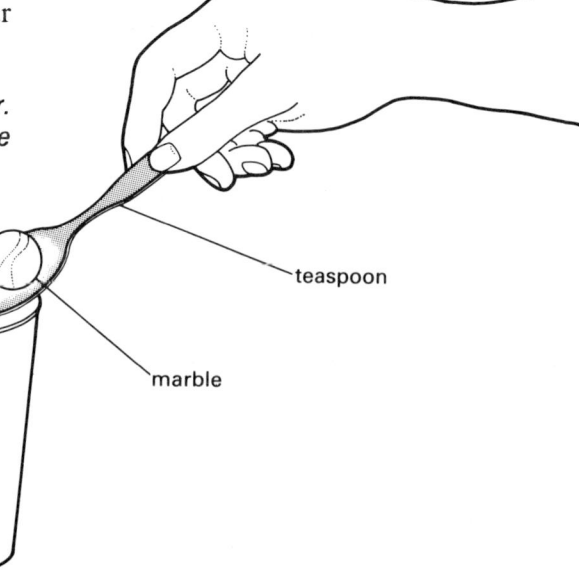

2 Make a cross between two individuals with the genotype *Aa* and *aa*. Put one red and one blue marble into one container, and two blue ones into the other. Without looking into the containers, use the spoons to remove one marble from each container. **Copy the grid of 16 squares (Figure 64) and write in square 1 the genotype of the first two marbles that were removed.** Return each marble to its container.

Fig. 64 *Filling in the grid.*

Remove one marble from each container

Red represents gene A and blue gene a. If a red and blue marble are removed write Aa in a square. Repeat this process to give 16 gene recombinations

Aa			

(a) **Repeat this process 16 times and write down the genotypes of the offspring each time you remove two marbles.** (Remember that if you look into the containers the pairing is not random.) *(1)*

(b) **What genotypes were obtained?** *(2)*

(c) **What was the genotypic ratio?** *(2)*

(d) **If the sample had been much bigger, what genotypic ratio would you expect to obtain? Show how you arrive at your answer.** *(3)*

3 Similarly, make a cross between two heterozygotes (*Aa* × *Aa*).

(a) **Draw another grid like Figure 64 and record your results.** *(1)*

(b) **What genotypes were obtained?** *(2)*

(c) **What was the genotypic ratio?** *(2)*

(d) **If the sammle had been much bigger, what genotypic ratio would you expect to have obtained? Show how you arrive at your answer.** *(3)*

Total 20 marks

Taking it further

1 What offspring genotypes do you get if *AA* is crossed with *aa*, and *aa* is crossed with *aa*?

49 The sense of taste

TIME
Investigation: 30–40 minutes

The tongue is a sense organ, providing the brain with information about the flavour of food in the mouth. Covering different regions of the tongue are sensory areas, containing groups of cells called taste buds. There are more taste buds in some parts of the tongue than in others. There are four different types of taste bud, each responding to one of the following flavours: (a) sweet, (2) salty, (3) sour, (4) bitter.

You will apply solutions of compounds with different flavours, and try to find out how different taste buds are distributed on the tongue. Additionally, you are asked to find out how a complex flavour, such as onion, is detected.

Investigation

Materials

- 2 cm³ sweet-tasting solution (A)
- 2 cm³ salty-tasting solution (B)
- 2 cm³ sour-tasting solution (C)
- 2 cm³ bitter-tasting solution (D)
- 1–2 g crushed onion, in a petri dish
- Five cotton buds
- Pocket mirror
- Glass, or paper cup containing distilled water
- Paper cup

Method

1 You have solutions that are sweet (A), salty (B), sour (C), and bitter (D). Put a cotton bud into each of the solutions. Copy Table 21.

2 Put out your tongue and look at it in the mirror. Imagine four horizontal lines drawn across it, as in Figure 65. Take the cotton bud in the sweet-tasting solution (A) and apply it across region 1 of your tongue. **Record in the table if the solution was tasted [✓] or not tasted [×]**. Rinse out your mouth with a little distilled water. Spit out the washings into the empty cup.

Apply the same test to regions 2, 3, and 4 of your tongue, and record your results in the table. Take in turn the salty (B), sour (C), and bitter (D) solutions. Test each one by applying it to regions 1, 2, 3, and 4 of your tongue. Rinse out your mouth after each test. **Complete the table.** (8)

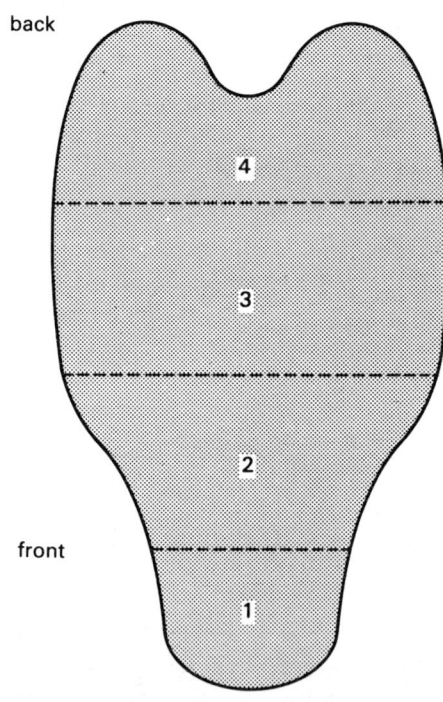

Fig. 65 *The human tongue, divided into four horizontal regions*

Table 21

Solution	Region of tongue			
	1	2	3	4
Sweet				
Salty				
Sour				
Bitter				

3 Dip a cotton bud into the crushed onion. Hold your nose and apply the dipped cotton bud across region 1 of your tongue. Rinse out your mouth a little distilled water. Similarly, repeat the test in regions 2, 3, and 4 of your tongue, holding your nose each time the cotton bud is applied. **What did you taste, and in which region(s)?** (2)

4 Repeat the test with crushed onion, but without holding your nose.

(a) **What difference did you notice?** (2)

(b) **Suggest a hypothesis to explain any difference.** (3)

5 **Why was it necessary to wash out your mouth with distilled water after each test?** (2)

6 **Some people believe there are taste buds under the tongue. How would you test this hypothesis?** (3)

Total 20 marks

Taking it further

1 Divide the tongue longitudinally into three regions. Use cotton buds dipped into the same solutions to test each region for taste. From your results try to make a map of those areas of the tongue responding to each solution.

2 A product called Canderel is sold as an artificial sweetner. Dissolve a tablet of this product in 10 cm³ of water, dip a cotton bud into it, and apply it to your tongue.
  (a) Is it tasted in the same region as sugar?
  (b) How does the taste compare with sucrose?
  (c) Is the product a reducing sugar, starch or protein (see Investigation 4)?
  (d) What are the advantages of using a product of this type to sweeten food?
  (e) Are there any disadvantages to using this product?

50 The skin as a sense organ

TIME
Investigation: 30–40 minutes

The skin is a sense organ, containing many sensory nerve-endings. These nerve-endings are of distinct types, each responding to a different stimulus. Those that respond to heat are distinct from those that respond to cold. Again, nerve-endings responding to touch are distinct from those responding to pressure. There are more of each type of nerve-ending in some parts of the skin than in others. As a result, sensations such as heat are most readily felt in regions where heat-sensitive nerve endings are most numerous. Some nerve-endings are superficial, positioned just beneath the epidermis. Others lie deeper, and are only reached by stimuli that can pass through most of the dermis. See Figure 66.

Fig. 66 *Sensory nerve-endings in skin*

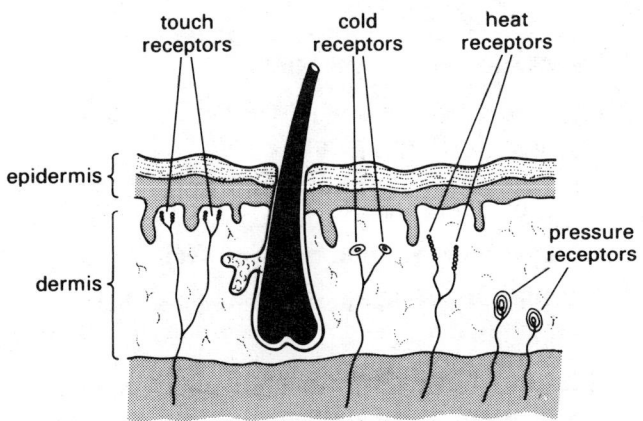

Investigation

Materials

- 5 cm³ ethanol
- Pencil
- Dry cotton wool

Method

1 Bare your left arm to the elbow, and roll up your sleeve. Take the dry cotton wool in your right hand. Close your eyes and stroke the cotton wool lightly over
   (i) the palm of your left hand;
   (ii) the back (dorsal side) of your left arm, and
   (iii) the front (ventral side) of your left arm.

(a) **Name the sensation that you feel.** (*1*)

(b) **Which type of nerve-ending is being stimulated?** (*1*)

(c) **Are these nerve endings superficial or deep?** (*1*)

(d) **In which of the three regions was the sensation most noticeable?** (*1*)

(e) **What conclusion can you draw about the number of nerve-endings in this region?** (*1*)

2 Rub the palm of your right hand vigorously backwards and forwards for 20–30 movements against (i) the palm of your left hand; (ii) the back of your left arm, (iii) the front of your left arm.

(a) **Name the sensation that you feel.** (*1*)

(b) **Which type of nerve-ending is being stimulated?** (*1*)

(c) **In which of the three regions was the sensation most noticeable?** (*1*)

3 Dip the cotton wool into ethanol. Apply ethanol to (i) the palm of your left hand; (ii) the back of your left arm, and (iii) the front of your left arm.

(a) **Name the sensation that you feel.** (*1*)

(b) **Which type of nerve-ending is being stimulated?** (*1*)

(c) **In which of the three regions was the sensation most noticeable?** (*1*)

4 Close your eyes. Press the point of the pencil into: (i) the palm of your left hand; (ii) the back of your left arm; (iii) the front of your left arm.

(a) **Name the sensation that you feel.** (*1*)

(b) **Which type of nerve-ending is being stimulated?** (*1*)

(c) **Are these nerve-endings more superficial or deeper than those responding to touch?** (*1*)

(d) **In which of the three regions was the sensation most noticeable?** (*1*)

5 Doctors sometimes use a small grid, similar to the one shown in Figure 67, to test the skin for responses to stimuli.
How would you use the grid to find out if someone had lost the sense of feeling (touch) in the palm of their hand? (*5*)

Total 20 marks

Fig. 67 *Grid for skin testing*

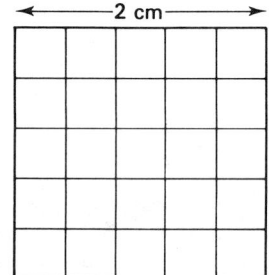

Taking it further

1 During vigorous exercise, when we get hot, sweat is produced by sweat glands in the skin. These, like sensory nerve-endings, are not evenly distributed throughout the skin, but are more numerous in some parts than in others.

Large self-adhesive wound dressings, 6 × 8 cm, can be stuck over parts of the skin to absorb the sweat produced during exercise. They are weighed before and after the exercise to find out how much sweat they have absorbed.

Design an experiment to find out if there are more sweat glands on the palm of your hand than on the sole of your foot.

(a) List the materials and apparatus for your experiment.
(b) Write out your method, listing instructions in the order they should be carried out.
(c) Draw a table, with headings, for your results.
(d) What assumptions have to be made?